PCR

ESSENTIAL DATA SERIES

Series Editors

D. Rickwood
Department of Biology, University of Essex,
Wivenhoe Park, Colchester, UK

B.D. Hames
Department of Biochemistry and Molecular Biology,
University of Leeds, Leeds, UK

Published titles
**Centrifugation
Gel Electrophoresis
Light Microscopy
Vectors
Human Cytogenetics
Animal Cells: culture and media
Cell and Molecular Biology
PCR**

Forthcoming titles
**Nucleic Acid Hybridization
Enzymes in Molecular Biology
Transcription Factors
Protein Purification
Immunoassays**

See final pages for full list of titles and order form

PCR
ESSENTIAL DATA

Edited by

C.R. Newton

Zeneca Pharmaceuticals, Macclesfield, UK

JOHN WILEY & SONS

Chichester · New York · Brisbane · Toronto · Singapore

Published in association with BIOS Scientific Publishers Limited

©1995 John Wiley & Sons Ltd, Baffins Lane, Chichester, West Sussex PO19 1UD, UK, tel (01243) 779777. Published in association with BIOS Scientific Publishers Ltd, 9 Newtec Place, Magdalen Road, Oxford OX4 1RE, UK.

All rights reserved. No part of this book may be reproduced by any means, or transmitted, or translated into a machine language without the written permission of the publisher.

British Library Cataloguing in Publication Data
A catalogue record for this book is available from the British Library.

ISBN 0 471 95222 2

Typeset by Marksbury Typesetting Ltd, Bath, UK
Printed and bound in UK by H. Charlesworth & Co. Ltd, Huddersfield, UK

The information contained within this book was obtained by BIOS Scientific Publishers Limited from sources believed to be reliable. However, while every effort has been made to ensure its accuracy, no responsibility for loss or injury occasioned to any person acting or refraining from action as a result of the information contained herein can be accepted by the publishers, authors or editors.

CONTENTS

Contributors xii
Abbreviations xiv
Preface xvii
Patents and license requirements xix
Safety note xx

1. Background to PCR and its application areas. C.R. Newton 1
Application areas 2

Tables
Quantitative analysis of PCR after 25 cycles 3
Technique references 4

2. Setting up a PCR laboratory. C.R. Newton 7
Consumables 8

Tables
Consumables 10
Important controls 11

3. Instruments. C.R. Newton 12

Tables
Thermal cyclers: I 13
Thermal cyclers: II 15
Thermal cyclers: III 18
Thermal cyclers: IV 20

4. Nucleic acid substrates. C.R. Newton 24
Examples of crude preparations for isolating DNA for PCR 25
Nucleic acid extraction kits 26

Contents

Tables

DNA yields from different human tissue sources	27
mRNA isolation kits	28
Total RNA isolation kits	29
DNA isolation kits	31
cDNA kits	32
Genomes and DNA substrates (micro-organisms)	33
Genomes and DNA substrates (plants)	34
Genomes and DNA substrates (animals)	35
Efficiencies	36

5. Thermostable DNA polymerases. C.R. Newton   37

DNA polymerases without 3' to 5' exonuclease activity (3' exo⁻)	39
DNA polymerases with 3' to 5' exonuclease activity (3' exo⁺)	43

Tables

Thermostable DNA polymerases	47
Properties of thermostable DNA polymerases	48

Figures and tables

Addition of biotin labels	58
Addition of digoxigenin to a synthetic oligonucleotide	60
Addition of DNP labels	60
Applied Biosystems fluorescent dyes	62
Applied Biosystems fluorescent dye phosphoramidites	63
Other 5' chemical labels	64
Monomer used to add phosphate to a synthetic oligonucleotide	65
Modified nucleoside phosphoramidite monomers	66
'PCR stoppers'	67
Absorption and emission wavelengths	69
Common chemical labels and their uses	70

8. Protocol optimization and reaction specificity.   72
S.J. Powell

Reaction components	72

6. Primers. C.R. Newton	**49**
Primer design	49
Calculating extinction coefficients for oligonucleotides and converting OD readings to micrograms and nanomoles	50
Calculating T_ms of oligonucleotides	52
Primer labeling	53
Tables	
Primer design	55
Micromolar extinction coefficient multiples	56
7. The synthesis of chemically labeled PCR primers. T. Brown and D.J.S. Brown	**57**
Oligonucleotide synthesis and purification	57
Chemical labels and labeled PCR primers	58
Suppliers of phosphoramidite monomers and other chemical labeling molecules	68
Thermal cycling	74
Nested PCR	77
Long PCR	77
Tables	
Optimization for amplification of fragments up to 5 kbp	79
Importance of optimization with respect to PCR applications	80
Buffer components and properties	81
PCR enhancers	82
Choice of polymerase to suit optimization criteria	83
Optimization properties of thermal cycling parameters	85
Polymerase combinations for long amplicons	86
9. Contamination avoidance. C.R. Newton	**87**
Tables	
Decontamination methods for PCR reaction mixtures	88

Contents

Decontamination methods for RT-PCR	90	Staining PCR products with EtBr	117
Decontamination methods for laboratory surfaces	92	AGE and PAGE preparation	118
		Resolution of DNA in agarose	120
10. Use of 5′ chemically labeled primers in PCR. J. Grzybowski, F. McPhillips, D.J.S. Brown and T. Brown	**93**	Resolution of DNA in polyacrylamide	120
		Types and applications of PAGE	120
		Factors affecting the polymerization of polyacrylamide gels	121
Modified primers containing nonradioactive labels	93	Parallel and perpendicular DGGE	122
Effect of modified oligonucleotide primers on PCR efficiency	95	Methods and applications of DNA blotting	123
		Commercially available nucleic acid blotting membranes	124
Figures and tables		Blotting procedures	125
The structure of PCR primer A	93	DNA transfer for Southern blotting	126
Addition of biotin to the 5′ end of a synthetic oligonucleotide	94	**12. Purification of PCR products.** C.R. Newton	**128**
Structure of synthetic oligonucleotide with DNP group added	95	*Tables*	
Multiple attachment of fluorescein to a synthetic oligonucleotide	96	Kits for DNA isolation	130
	97	Molecular weight cut-offs for DNA purification	131

Attachment of a cholesterol moiety to a synthetic oligonucleotide	96	**13. Cloning PCR products.** C.R. Newton	**132**
Structure of a synthetic oligonucleotide with digoxigenin attached	96	*Figures and tables*	
Effect of modified primers on PCR efficiency	98	pGEM-T vector, promoter and multiple cloning sites	132
		pCR^{TM}II vector, promoter and multiple cloning sites	133
11. Detection of PCR products. L.J. MacCallum	**99**	pCR^{TM}III vector, promoter and multiple cloning sites	134
Homogeneous detection of PCR products	99	pCITE vector, promoter and multiple cloning sites	135
Gel detection of PCR products	102	pT7 Blue vector, promoter and multiple cloning sites	134
Membrane detection of PCR products	105		
Detecting labeled PCR products	106	pTOPE vector, promoter and multiple cloning sites	135
Tables		pMOS*Blue* vector, promoter and multiple cloning sites	136
Techniques for detecting PCR products	109	pCR Script^{TM} vector, promoter and multiple cloning sites	137
Commercially available amplicon DNA detection kits	112		
Fluorescence of EtBr in UV light	115		
Loading buffers for gel electrophoresis	115		
Buffers for gel electrophoresis systems	116		

Contents

pNoTA vector, multiple cloning sites	136
pCR-TRAP vector	136
pDIRECT vector, promoter and multiple cloning sites	137
pAMP1 vector, promoter and multiple cloning sites	138
pAMP10 vector, promoter and multiple cloning sites	138
pAMP18 vector, multiple cloning sites	138
pAMP19 vector, multiple cloning sites	138
Features of PCR cloning vectors: I	139
Features of PCR cloning vectors: II	140
Attributes of PCR cloning systems	141

14. Miscellaneous data. C.R. Newton — **143**

Restriction endonuclease activity in PCR buffer (P. Eastlake, M. Pinkney and P. Gorringe)	143
Cleavage close to the end of DNA fragments: I	149
Cleavage close to the end of DNA fragments: II	155
Isotopes in PCR	155
Decay table for ^{32}P	156
Decay table for ^{33}P	157
Decay table for ^{35}S	156
The genetic code	157
Degeneracy codes	158
Amino acid abbreviations	158
Codon degeneracy	158
Restriction site introduction via silent mutations	159
Taq DNA polymerase misincorporation preference	161
Metric conversions	161
Other DNA data	161
RT-PCR kits	162

Data for radioisotopes commonly used in PCR	145
Genetic code, DNA degeneracy code	145
Amino acid abbreviations and codon degeneracy and restriction site introduction via silent mutations	145
Misincorporation rates by *Taq* DNA polymerase and other DNA	144
Metric conversions and conversion data	144
Commercially available kits and specialized reagents for PCR and associated techniques	144
Tables	
Restriction enzyme activity in PCR buffer	147

15. Troubleshooting. C.R. Newton	**163**
Tables	
PCR troubleshooting guide	163
16. Manufacturers and suppliers. C.R. Newton	**168**
References	**180**
Further reading	**207**
Appendix	**209**
Trade names and registered trademarks	209
Index	**213**

CONTRIBUTORS

D.J.S. Brown
OSWEL DNA Service, Medical and Biological Science Building, University of Southampton, Bolderwood, Bassett Crescent East, Southampton SO16 7TX, UK

T. Brown
OSWEL DNA Service, Medical and Biological Science Building, University of Southampton, Bolderwood, Bassett Crescent East, Southampton SO16 7TX, UK

P. Eastlake
NBL Gene Sciences Ltd, South Nelson Road, Cramlington, Northumberland NE23 9HL, UK

P. Gorringe
NBL Gene Sciences Ltd, South Nelson Road, Cramlington, Northumberland NE23 9HL, UK

L.J. MacCallum
Zeneca Pharmaceuticals, Mereside, Alderley Park, Macclesfield, Cheshire SK10 4TG, UK

F. McPhillips
Department of Chemistry, University of Edinburgh, King's Building, West Mains Rd, Edinburgh EH93 3JJ, UK

C.R. Newton
Zeneca Pharmaceuticals, Mereside, Alderley Park, Macclesfield, Cheshire SK10 4TG, UK

M. Pinkney
NBL Gene Sciences Ltd, South Nelson Road, Cramlington, Northumberland NE23 9HL, UK

J. Grzybowski
Department of Biochemistry, University of Liverpool, PO Box 147, Liverpool L69 3BX, UK

S.J. Powell
Zeneca Pharmaceuticals, Mereside, Alderley Park, Macclesfield, Cheshire SK10 4TG, UK

ABBREVIATIONS

A	adenine	EGTA	ethylene glycol-bis(β-aminoethyl ether) N,N,N',N'-tetraacetic acid
Ac	acetate		
AGE	agarose gel electrophoresis	ELISA	enzyme-linked immunosorbent assay
AP	arbitrarily primed (PCR), alkaline phosphatase	E-PCR	expression-PCR
APS	ammonium persulfate	EtBr	ethidium bromide
ARMS	amplification refractory mutation system	FPLC	fast protein liquid chromatography
ASA	allele-specific amplification	G	guanine
ASO	allele-specific oligonucleotide	GAWTS	gene amplification with transcript sequencing
ASP	allele-specific PCR	GSP	gene-specific primer
ATP	adenosine triphosphate	HLA	human leukocyte antigen
bp	base pair	HPLC	high pressure liquid chromatography
BSA	bovine serum albumin	HRP	horseradish peroxidase
C	cytosine	kbp	kilobase pair
cDNA	complementary deoxyribonucleic acid	LMP	low melting point
COP	competitive oligonucleotide priming	mol. wt	molecular weight
dATP	deoxyadenosine triphosphate	mRNA	messenger RNA

Abbreviations

dCTP	deoxycytidine triphosphate	nt	nucleotides
ddNTP	dideoxynucleoside triphosphate	OD	optical density
DDRT	differential display reverse transcriptase (PCR)	PAGE	polyacrylamide gel electrophoresis
DEPC	diethylpyrocarbonate	PASA	PCR amplification of specific alleles
DEVIATS	double-ended vectorette incorporating alternative transcription sites	PBS	phosphate-buffered saline
		PEG	polyethylene glycol
DGGE	denaturing gradient gel electrophoresis	PEST	primer extension sequence test
dGTP	deoxyguanosine triphosphate	RACE	rapid amplification of cDNA ends
dIMP	deoxyinosine monophosphate	RAPD	random amplified polymorphic DNA
DMSO	dimethylsulfoxide	RAWIT	RNA amplification with *in vitro* translation
dNMP	deoxynucleoside monophosphate	RAWTS	RNA amplification with transcript sequencing
DNP	dinitrophenyl	RFLP	restriction fragment length polymorphism
dNTP	deoxynucleoside triphosphate	RNase	ribonuclease
dsDNA	double-stranded DNA	rRNA	ribosomal ribonucleic acid
dTMP	deoxythymidine monophosphate	RT	reverse transcriptase
DTT	dithiothreitol	SDS	sodium dodecyl sulfate
dTTP	deoxythymidine triphosphate	SNuPE	single nucleotide primer extension
dUMP	deoxyuridine monophosphate	SOE	splicing by overlap extension
EC-PCR	expression cassette-PCR	SSB	single-stranded binding protein
EDTA	ethylenediaminetetraacetic acid	SSCP	single-strand conformation polymorphism

ssDNA	single-stranded DNA	TMAC	tetramethylammonium chloride
T	thymine	U	uracil
TA	thymine, adenine	UDG	uracil-DNA glycosylase
TAPS	Tris-(hydroxymethyl)-methyl-amino-propanesulfonic acid	UV	ultraviolet
		v/v	volume/volume
TBR	Tris (2,2'-bipyridine) ruthenium (II) chelate	w/v	weight/volume
TEMED	N,N,N',N'-tetramethylethylenediamine	YAC	yeast artificial chromosome
T_m	melting temperature		

PREFACE

In the few years since the polymerase chain reaction (PCR) has been introduced to the research community it has become an indispensable tool for many areas of investigation. Not only does it now play a major role in many scientific disciplines, but it is also the bedrock of many specific techniques within those disciplines. There are frequent instances of PCR techniques having been developed and passed into the service laboratory environment. These service laboratories are providing an extensive range of diagnostic tests covering veterinary, environmental, agricultural, medical and forensic applications. A significant industry has also grown to provide the PCR user. This has led to the rapid development of instrumentation, reagents, reagent kits, enzymes, consumables and discrete services such as custom oligonucleotide synthesis and DNA sequencing. One of the aims of this book is therefore that it should serve as a guide for those about to purchase equipment, consumables, reagents, etc., such that the final decision matches the need for a specific PCR application.

Despite the introduction and broad spectrum of well-developed reagent kits, PCR still fails from time to time. The other aim of this book is therefore to provide the relevant information, directly or through appropriate literature citations, to answer *practical* questions regarding the PCR technique. This is backed up by a comprehensive troubleshooting guide. Towards achieving this aim, I hope that the reader will find this book instructive on an ongoing basis and as a continuing source of reference.

Trade names and registered trademarks appearing throughout the text are the properties of the organizations as shown in the Appendix. Further reading is suggested from which a greater understanding of specific PCR applications may be

Preface

gleaned. A comprehensive listing is provided in Chapter 16 of suppliers of products discussed in the text.

This book is, as the series name states, a compendium of essential information for active researchers who are already using PCR or who wish to do so. A companion volume [1] provides the complementary information by way of describing in detail the wide range of PCR-based techniques and applications now available. It is particularly useful to those who are about to embark on PCR for the first time or are considering a new aspect of PCR methodology.

C.R. Newton

PATENTS AND LICENSE REQUIREMENTS

The reader is hereby notified that the purchase of this book does not convey any license or authorization to practice PCR under any patents owned by Hoffmann–La Roche Inc. and its affiliates. Neither is this book intended to suggest that scientists use reagents for PCR that are not licensed under these patents for such use.

The basic PCR process is covered by US patent numbers 4,683,202; 4,683,195; 4,800,159 and 4,965,188 owned by Hoffmann–La Roche Inc. and other pending and issued patents in non-US countries owned by F. Hoffmann–La Roche A.G. The Perkin–Elmer Corporation is the exclusive distributor of Roche's PCR products in certain fields and has the limited rights under the above PCR patents in those fields.

Taq DNA polymerase and AmpliTaq® DNA polymerase are covered by US patent numbers 4,889,818 and 5,079,352 owned by Hoffmann–La Roche Inc. By purchasing Perkin–Elmer PCR reagents for use with Perkin-Elmer-authorized thermal cyclers, the purchasers obtain a limited license for research and development purposes under US patents 4,683,202; 4,683,195; 4,965,188 and 5,075,216 or their foreign counterparts. No right to offer or perform commercial services of any kind using PCR is granted unless the purchaser obtains a further license by contacting either Perkin–Elmer or Hoffmann–La Roche.

At the time of going to press, the companies licensed to provide research reagents for PCR comprise Amersham International, Boehringer Mannheim, Takara-Schuzo Co., Life Technologies, Inc. (GIBCO-BRL) and Stratagene.

Further information on purchasing licenses to practice the PCR process may be obtained by contacting the Director of Licensing at The Perkin–Elmer Corporation, 850 Lincoln Centre Drive – M/S 408, Foster City, CA 94404, USA or at Roche Molecular Systems, Inc., 1145 Atlantic Avenue, Alameda, CA 94501, USA.

SAFETY NOTE

Data sheets supplied with reagents should be read and any hazard label warnings should be noted. Local rules and suppliers' recommendations pertaining to the safe use and handling of chemicals and radioisotopes should also be observed. The use of ^{35}S-labeled nucleotides in PCR reactions has led to reports of contaminated thermocycler sample blocks. If possible, alternative radiolabels such as ^{33}P should be substituted or nonisotopic labels considered. If the use of ^{35}S is unavoidable, then the thermocycler should be operated within a fume hood and PCR reactions should be overlaid with wax or mineral oil to minimize the release of volatile ^{35}S compounds. Biological samples are able potentially to transmit hepatitis and other infectious diseases, so they should be handled with the appropriate precautions, and it is recommended that operations involving the mixing of blood, or cells derived from blood, be performed in a safety cabinet. Gloves, safety spectacles and laboratory coats should also be worn. Never pipette by mouth. While every care has been taken to ensure that the experimental details discussed in this book meet all relevant safety requirements, the editor and contributors accept no liability for any loss or injury howsoever caused.

Chapter 1 A BACKGROUND TO PCR AND ITS APPLICATION –

C.R. Newton

The polymerase chain reaction [1–4] is an *in vitro* nucleic acid amplification method. The reaction comprises repeated thermal cycling of the reaction mixture. The thermal cycling serves to dissociate the products of the previous thermal cycle then allow the association of these dissociated products with further reaction starting materials for another phase of synthesis. Specifically, substrate DNA is denatured at a high temperature to give single-stranded (template) molecules. This is followed by short oligonucleotide primers (amplimers) annealing to specific, nucleotide-sequence-defined regions of the template at a lower temperature. The positions where the amplimers anneal to the template define the 'target'. The thermal cycle is concluded by the amplimers being enzymatically extended by the coupling of appropriately base-paired deoxynucleoside triphosphates (dNTPs) on the template at an intermediate temperature, thus producing another double-stranded DNA (dsDNA) copy of the original target.

Each PCR cycle comprises a set of time- and temperature-controlled incubations. The function of each incubation is to:

1. Denature the target nucleic acid at a temperature in the region of 94°C;
2. Anneal amplimers at a temperature dependent on their calculated annealing temperature (usually in the range 30–60°C);
3. Extend the primers by the thermostable DNA polymerase catalyzed addition of nucleotides to the 3' end of each primer at a temperature of about 72°C.

If each PCR cycle were 100% efficient, each cycle would give rise to a doubling of the number of copies of the original

Background and Application

target sequence, since the product of one PCR cycle becomes additional substrate for the next. The PCR process can therefore give rise to millions of copies of the target sequence, enabling trace amounts to be detected against a background of a complex mixture of sequences. PCR may be used to analyze or determine the presence of specific DNA targets or, after a reverse transcription step, RNA sequences can also be amplified as DNA copies (see Chapter 14, *Table 16*).

Table 1 illustrates how reaction components alter over the course of a PCR.

1 Application areas

Selected references and reviews covering the broad application areas of PCR include: PCR, general [1–8]; basic molecular biology [6–10]; DNA sequencing [11, 12]; evolutionary studies [13–15]; forensic science [14, 16]; genetic analysis [14, 17–20]; population genetics [13, 14, 21]; medical diagnostics [5, 14, 19, 22–28]; veterinary diagnostics [21]; clinical immunology [9, 14, 24, 29]; agricultural biotechnology [30]; food testing [31–33] and environmental testing [34–36]. Citations are expanded further in *Table 2* to cover specific adaptations of PCR that have been developed in response to specific goals within a given application.

Table 1. Quantitative analysis of PCR after 25 cycles assuming 10^5-fold amplification (reproduced courtesy of Applied Biosystems)

	Before PCR				After PCR			
	Weight	Moles	Molarity	Molecules	Weight	Moles	Molarity	Molecules
Template[a]	1 ng	3.10×10^{-17}	3.10×10^{-13}	1.86×10^{7}	1 ng	3.00×10^{-17}	3.00×10^{-13}	1.81×10^{7}
Target[b]	10 pg	3.00×10^{-17}	3.00×10^{-13}	1.81×10^{7}	1 μg	3.00×10^{-12}	3.00×10^{-8}	1.81×10^{12}
Primers[c]	1623 ng	2.00×10^{-10}	2.00×10^{-6}	1.20×10^{14}	1574 ng	1.94×10^{-10}	1.94×10^{-6}	1.17×10^{14}
dNTPs[d]	39 μg	8.00×10^{-8}	8.00×10^{-4}	4.82×10^{16}	37 μg	7.70×10^{-8}	7.70×10^{-4}	4.64×10^{16}
Magnesium ion[e]	3.6 μg	1.50×10^{-7}	1.50×10^{-3}	9.03×10^{16}	3.6 μg	1.50×10^{-7}	1.50×10^{-3}	9.03×10^{16}
Taq DNA polymerase[f]	12.5 μg	1.33×10^{-13}	1.33×10^{-9}	8.01×10^{10}	12.5 μg	1.33×10^{-13}	1.33×10^{-9}	8.01×10^{10}

[a]Bacteriophage lambda (template dsDNA = 48 500 bp).
[b]Target is 500 bp.
[c]1 μM (each) primers, 25-mers.
[d]200 μM (each) dNTPs; total [dNTPs] = 0.8 mM (average mol.wt of a dNTP is 487 Da; average mol.wt of a dNMP is 325 Da).
[e]Total [MgCl$_2$] = 1.5 mM; free [MgCl$_2$] = 0.7 mM.
[f]2.5 Units *Taq* DNA polymerase per 100 μl; polymerase activity = 250 000 units mg^{-1}; enzyme half-life not considered.

Table 2. Technique references

Application	Technique	References
General	Hot-start	37, 38
	Nested PCR	39, 40
	Contamination avoidance	41–55
	Multiplex PCR	56–74
	Long PCR	75–81
Cloning	Restriction site incorporation	82, 83
	Blunt end	83–85
	SOE (recombinant PCR)	86–91
	TA vectors	92–94
	Half-site generation	95
	LIC	96–102
	Ligation anchored PCR	103
	RACE	104–109
	RT-PCR	43, 110–120
	DDRT-PCR (mRNA differential display)[a]	121, 122
	In situ RT-PCR[a]	123
	Gene synthesis	124–134
	Mutagenesis	135–171
	Expression cloning	172
	Expression PCR[a]	173–175
Sequencing	Direct	176–190
	Asymmetric PCR	188, 191, 192

	Solid phase sequencing	184–186, 189, 190
	Cycle sequencing	193
	GAWTS	194, 195
	RAWTS	196
Archaeology		13–15, 197–204
Forensics	HLA typing	202–213
	Microsatellites	214–225
Unknown sequence amplification	Vectorette (bubble) PCR and DEVIATS	226–235
	Inverse PCR	236–243
Clinical pathology	Virus detection	61, 67, 73, 244–250
	Bacteria detection	32, 33, 72, 74, 251–257
	Histological specimen analysis	218, 220, 258–268
	In situ PCR	265–275
	RT *in situ* PCR	276–280
	Diagnosing residual disease after cancer therapy	23, 281–287
Genetic diagnosis	ARMS (ASA, PASA, ASP)	209–212, 238, 281, 288–303
	Multiplex ARMS	62, 304–307
	Multiplex PCR	56, 57, 64, 66, 308–316
	Dot blot	317
	Reverse dot blot	213, 318–320
	COP	321, 322
	PEST	323
	Taq 5′–3′ exo	324

Continued

5   *Background and Application*

Table 2. Technique references, *continued*

Application	Technique	References
	Introduction of restriction sites	325–331
	RFLP	69, 332, 333
	SNuPE	334–337
	Preimplantation diagnosis	65, 70, 338–343
Characterizing unknown mutations	DGGE	344–347
	SSCP	299, 348–358
	Chemical cleavage	358–364
Fingerprinting/population analysis	RAPD (AP-PCR)	365–380
	Microsatellites	66, 214–225, 381–387
	HLA typing	202–213
Genome analysis	End mapping of cosmids/YACs	229, 388–399
	Chromosome mapping	226, 232, 234, 400
	Alu-PCR	401–412
		99, 388, 390, 391, 410, 413–429
Quantitative PCR	For DNA	430–436
	For RNA	248, 249, 314, 437–440
		111–119, 441–445

[a]This technique does not actually require cloning the amplicon.
For abbreviations see list on p. xiv.

Chapter 2 **SETTING UP A PCR LABORATORY** – C.R. Newton

The very nature of the exponential amplification by the PCR poses a major problem, that of contamination of samples intended for amplification by prior PCR amplicons. This type of contamination is otherwise known as 'PCR product carry-over'. Although there are biochemical means of reducing or preventing this (see Chapter 9), the risk can be reduced dramatically by the appropriate design and separation of the laboratory areas where the PCRs and associated procedures are performed [1]. Physical separation of different work areas also reduces the chance of accidental PCR product carry-over since it is likely to separate the pipettes used for each operation. To illustrate this, 0.1 μl of a PCR reaction mixture may contain 3×10^9 amplicon molecules; 1 μg of human genomic DNA, in contrast, represents about 3×10^5 copies of a single-copy gene. The laboratory set up should therefore be designed to contain amplified DNA and so prevent it from ever coming into contact with samples or PCR reagents prior to amplification. The laboratory should ideally have three separate areas:

1. Area 1, sample handling and DNA extraction;
2. Area 2, setting up the PCRs;
3. Area 3, performing the PCRs and carrying out their analysis.

The separation of areas 1 and 2 serves to avoid transfer of exogenous DNA into area 2; a critical requirement for archaeological, forensic and clinical diagnostic applications. Any air flow should also be from the first and second areas to the third or completely independent to avoid transmission of aerosols from area 3 to areas 1 and 2. Pipettors should be dedicated to the specific areas and the pipettors used in area 2 for adding DNA to the PCR reaction mixture should be positive displacement with disposable tips and plungers. Ideally, area 2 should be equipped with a microbiological safety cabinet with an ultraviolet (UV) source. The DNA addition to the PCR reaction mixture should be performed in the safety cabinet since the germicidal UV irradiation will damage DNA left on exposed surfaces, rendering it unamplifiable.

1 Consumables

In many instances the choice of reaction tube or other vessels will be dictated by the choice of thermocycler (see Chapter 3). In general, it is recommended that where a thermocycler manufacturer also supplies reaction tubes, etc., then the consumables are acquired from the same source. This is because of vessel/heating block optimization during design and manufacture; in particular, tube fit which is critical for efficient thermal transfer. Some of the major suppliers of PCR laboratory consumables are listed in *Table 1*.

4. Prepare and store reagents in an area free of amplified DNA (e.g. area 2 above or in a separate preparation area).
5. Synthesize and purify amplimers in an area free of amplified DNA (e.g. area 2 above or in a separate preparation area).
6. To aid the tracing of any contamination, keep records of batches of reagents.
7. Wear disposable gloves and change them frequently and always when entering area 1 or area 2 from area 3.
8. Avoid splashes, e.g. when opening microfuge tubes; it is not necessary to blow the final residue from a pipettor tip, the negligible volume compared to a PCR volume is not worth the risk of generating an aerosol.
9. Premix reagents before dividing into aliquots and combining with sample DNA to minimize the number of reagent transfers.
10. Add sample DNA to the reaction mixture last to reduce the possibility of cross-contamination from one reaction to the next.

1.1 Post-PCR workup consumables
See Chapter 11 for detection and Chapter 12 for purification of PCR products.

Laboratory practice
1. Autoclave solutions for sample preparation and PCR and deionized or distilled water (do not autoclave template DNA, amplimers, DNA polymerase or dNTPs).
2. Autoclave disposable pipettor tips and all other disposable plasticware unless these have been supplied sterile.
3. Divide reagents and fresh preparations of solutions into aliquots to minimize the number of samplings.
11. Include a negative control with no sample DNA but all of the other PCR reagents. Prepare this last to represent the total reagents handled.
12. Choose controls carefully (*Table 2*).
13. If there is any doubt about a critical result, repeat the experiment – the chances of a sporadic event occurring the same way twice are low.
14. Avoid cross-contamination via electrophoresis equipment by washing gel combs and casting trays in 1 M HCl.
15. When excising bands from gels, use a new scalpel blade for each band.
16. Cover the transilluminator with a fresh sheet of plastic wrap for every gel.

Setting up a PCR Laboratory

Table 1. Consumables

0.2 ml tubes[a]	0.2 ml tubes[a] (thin wall)	0.5 ml tubes[a]	0.5 ml tubes[a] (thin wall)	Microtiter plates (96-well)	Microtiter plates (384-well)	Plugged pipette tips	+ve displacement pipettes and tips	Wax beads
ABI[b]	ADB[b]	ABI[b]	ABI[b]	ABI[b, c]	Anachem	ABI[b]	Anachem	ABI[b]
Costar	Appligene	Bio-Rad	ADB[b]	Appligene		Anachem	BCL[b]	ADB[b]
MJ Research	Bio-Rad	Camlab	Anachem	Biometra		Appligene	Costar	Flowgen
RSC[b]	Costar	Costar	Appligene	Costar		Bio-Rad	Eppendorf	NBL[b, f]
	Genhunter	Life Sciences	Costar	GIBCO-BRL		Costar	Gilson	
	Life Sciences	MJ Research	Genhunter	Hybaid		GIBCO-BRL	Hybaid	
	NBL[b]	PGC Scientific	ISS[b]	MJ Research		ISS[b]	Labsystems	
	RSC[b]	RSC[b]	ISS[b, d]	NBL[b]		Labsystems	Life Sciences	
	Techne	Sarstedt	NBL[b, d]	RSC[e]		Life Sciences	NBL[b]	
		Sorenson	PGC Scientific	Stuart Scientific[g]		MBP[b]	Techgen	
		Techne	RSC[b]	Techne		NBL[b]		
			Sorenson			PGC Scientific		
						Promega		
						Rainin		
						RSC[b]		
						Sorenson		

[a] Nominal. [b] Abbreviations: ABI, Applied Biosystems; ADB, Advanced Biotechnologies; BCL, Boehringer; ISS, Integrated Separation Systems; MBP, Molecular Bio-Products Inc.; NBL, NBL Gene Sciences; RSC, Robbins Scientific. [c] Rack and tubes (0.2 ml) in 96-well format. [d] Oil-free tubes available comprising internal shaft above reaction mixture. [e] Both trays and racks with tubes as in [c]. [f] Applied molten. [g] 32 in 4 x 8 format.

Table 2. Important controls

Control for	Method
DNA-free status of reagents and solutions	PCR in the absence of exogenously added DNA
Absence of DNA contamination at the detection limit of the PCR conditions – occurrence of sporadic contamination	Multiple PCRs in the absence of exogenously added DNA
Completeness of PCR mixture (checks whether essential components are missing or degraded)	PCR with positive control DNA
Sensitivity and efficiency	PCR with only sufficient positive control DNA to amplify weakly but consistently
Reaction specificity	Use negative and positive controls to check for spurious background bands
Set up, thermocycling, reaction sensitivity	Use negative and positive controls to check that PCR parameters are suitable and that expected product yields are obtained
Presence of 'amplifiable' DNA in the sample	PCR with primers for alternative target
Inhibition from endogenous substances in the sample	Spike control reaction with a known amplifiable target and its respective primers
Result plausibility	Repeat PCRs with primers from an independent but related sequence
Reliability of results	Perform PCR at least twice

Setting up a PCR Laboratory

Chapter 3 INSTRUMENTS – C.R. Newton

The development of the instrumentation to allow the automation of temperature cycling has undoubtedly contributed to the success of the PCR procedure. Thermal cycling parameters are critical to a successful PCR. These are listed as a guide to the important steps in the temperature and time profile of thermal cycling;

1. Denaturation;
2. Annealing of primers;
3. Extension of the primers;
4. Cycle number; and
5. Ramp times (the time taken to change from one temperature to the next in the thermal cycler).

Thermal cycling devices with a range of levels of sophistication are now available from many manufacturers and these have easy to use software for fully automated operation, cations and cycle sequencing or microscope slides for *in situ* PCR. Some machines have interchangeable sample formats, making these particularly versatile for different applications.

Tables 1–4 compare commercially available thermal cyclers and examine a variety of variables to help the prospective purchaser in the choice of instrument. It should also be noted that some manufacturers will supply custom-designed sample blocks or racks. The addresses, telephone numbers and facsimile numbers of the suppliers listed in *Tables 1–4* can be found in Chapter 16 for custom manufacture inquiries.

A further useful instrument is the self-contained, battery-operated temperature verification system available in either the 96-well microtiter or 0.5 ml microtube format from

together with accurate and reproducible temperature control for all of the samples in the cycler. Some of the instruments are able to accommodate alternative reaction formats such as tubes, 96-well microtiter arrays for high throughput applications. Applied Biosystems. These have been designed specifically for the GeneAmp PCR instrument systems to validate temperature calibration and uniformity, and comprise a hand-held meter and platinum resistance probe.

Table 1. Thermal cyclers: I

Supplier/manufacturer	Product name	Temperature control precision (°C)	Temperature accuracy (°C)	Programmable ramp rates	Temperature range (°C)
Applied Biosystems	GeneAmp PCR system 2400	+/- 0.5 above 35°C	+/- 0.75	Yes	4-99.9
Applied Biosystems	GeneAmp PCR system 9600	+/- 0.5 above 50°C	+/- 0.75	Yes	4-99.9
Applied Biosystems	DNA Thermal Cycler 480	+/- 0.5 above 35°C	+/- 1.0 above 33°C	Yes	-5-100
Applied Biosystems	DNA Thermal Cycler	+/- 1.0 above 35°C	+/- 1.0 above 33°C	No	0-99
Applied Biosystems	GeneAmp *in situ* PCR 1000	+/- 0.5 above 35°C	+/- 1.0 above 35°C	Yes	-5-99
Appligene	Crocodile II	+/- 0.1	+/- 1.0	Yes	5 above ambient-99
Appligene	Crocodile III	+/- 0.1	+/- 1.0	Yes	5 above ambient-99
Biometra	TRIO-Thermoblock	+/- 0.05	+/- 0.5	Yes	4-100
Biometra	UNO-Thermoblock	+/- 0.05	+/- 0.5	Yes	4-100
Biometra	Personal Cycler	+/- 0.05	+/- 0.5	Yes	4-100
BioTherm	Bio-Oven III	+/- 0.4	+/- 0.5	Yes	Ambient-125
Camlab	Thermojet +	+/- 0.1	+/- 0.5	Yes	-5-105
Eppendorf	Mastercycler 5330	+/- 0.2	+/- 0.8	No	4-96
Ericomp	SingleBlock System	+/- 0.1	+/- 0.5	Yes	Ambient-100
Ericomp	TwinBlock System	+/- 0.1	+/- 0.5	Yes	Ambient-100

Continued

Table 1. Thermal cyclers: I, *continued*

Supplier/ manufacturer	Product name	Temperature control precision (°C)	Temperature accuracy (°C)	Programmable ramp rates	Temperature range (°C)
Ericomp	DeltaCycler I (SingleBlock Peltier)	+/- 0.1	+/- 0.5	Yes	4–100
Ericomp	DeltaCycler II (TwinBlock Peltier)	+/- 0.1	+/- 0.5	Yes	4–100
Ericomp	PowerBlock System	+/- 0.1	+/- 0.4	Yes	4–100
Ericomp	SingleStar System	+/- 0.1	+/- 1.0	Yes	Ambient–100
Ericomp	TwinStar System	+/- 0.1	+/- 1.0	Yes	Ambient–100
GL Applied Research	GTC-2	+/- 0.2	+/- 0.2	No	0–99
Grant	Autogene II	+/- 0.5	+/- 0.5	No	4–99
Hybaid	Omnigene	+/- 1.0	+/- 0.5 (+/- 1.0 for *in situ* block)	Yes	5 above ambient–99
Hybaid	Omnislide	+/- 1.0	+/- 1.0	Yes	5 above ambient–99
Hybaid	Thermal reactor	+/- 1.0	+/- 1.0	Yes	5 above ambient–99
Idaho Technology	1605 Air Thermo-Cycler	+/- 0.5	+/- 0.25	Yes	5 above ambient–100
Integrated Separation Systems	Pro-Oven I	+/- 1.0	+/- 0.5	Yes	Ambient–125
Integrated Separation Systems	Pro-Oven III	+/- 1.0	+/- 0.5	Yes	Ambient–125
MJ Research	PTC-100	+/- 0.5	+/- 0.5	Yes	0–100
MJ Research	PTC-150 MiniCycler	+/- 0.4	+/- 0.3	Yes	−9–105
Stratagene	RoboCycler 40	+/- 0.1	+/- 0.5	No	6–99
Stratagene	RoboCycler Gradient 40	+/- 0.1 (A)	+/- 0.5	No	6–99

Stuart Scientific	SPCR1 GENE-TECH	+/- 0.5		Yes	4-99
Stuart Scientific	SPCR3 MINI-GENE	+/- 0.2		Yes	4-99
Techne	Cyclogene	+/- 1.0		Yes	4-99
Techne	Gene E	+/- 1.0		Yes	10 above ambient-99
Techne	PHC-1A	+/- 1.0		No	4-99
Techne	PHC-2	+/- 1.0		Yes	4-99
Techne	MW-1A	+/- 1.0		No	4-99
Techne	MW-2	+/- 1.0		Yes	4-99

A, Well to well gradient temperature uniformity of +/- 0.25°C across gradient block.

Table 2. Thermal cyclers: II

Supplier/ manufacturer	Product name	Cooling method	Concurrent operation of different programs	Number of programs	Number of segments per program
Applied Biosystems	GeneAmp PCR system 2400	Internal refrigeration	No	68	Unlimited
Applied Biosystems	GeneAmp PCR system 9600	Internal refrigeration	No	150	9
Applied Biosystems	DNA Thermal Cycler 480	Internal refrigeration	No	93	16
Applied Biosystems	DNA Thermal Cycler	Internal refrigeration	No	99	16
Applied Biosystems	GeneAmp *in situ* PCR 1000	Internal refrigeration	No	93	16
Appligene	Crocodile II	n.i.	No	27	8
Appligene	Crocodile III	n.i.	No	27	8
Biometra	TRIO-Thermoblock	Peltier	Yes, 3	30	12
Biometra	UNO-Thermoblock	Peltier	No	30	12

Continued

Table 2. Thermal cyclers: II, *continued*

Supplier/ manufacturer	Product name	Cooling method	Concurrent operation of different programs	Number of programs	Number of segments per program
Biometra	Personal Cycler	Peltier	No	99	99
BioTherm	Bio-Oven III	Air	No	99 (+ 99 profiles)	20 (+ 20 per profile)
Camlab	Thermojet +	Peltier	Yes, via smart card	16 + 2 per smart card	10
Eppendorf	Mastercycler 5330	Peltier	Yes	100	Unlimited
Ericomp	SingleBlock System	Water	No	99	99
Ericomp	TwinBlock System	Water	Yes	99	99
Ericomp	DeltaCycler I (SingleBlock Peltier)	Peltier	No	99	99
Ericomp	DeltaCycler II (TwinBlock Peltier)	Peltier	Yes	99	99
Ericomp	PowerBlock System	Peltier	Yes	99	99
Ericomp	SingleStar System	Forced air	No	99	99
Ericomp	TwinStar System	Forced air	Yes	99	99
GL Applied Research	GTC-2	Internal refrigeration	No	99	9
Grant	Autogene II	Water	No	50	7
Hybaid	Omnigene	Fan-driven air	Yes	18 + 36 per smart card	100
Hybaid	Omnislide	Fan-driven air	No	18 + 36 per smart card	100

Hybaid	Thermal reactor	Fan-driven air	No	25	80
Idaho Technology	1605 Air Thermo-Cycler	Fan-driven air	No	6	3
Integrated Separation Systems	Pro-Oven I	Fan-driven air	No	99	20
Integrated Separation Systems	Pro-Oven III	Fan-driven air	No	99	20
MJ Research	PTC-100	Peltier	No	200	100
MJ Research	PTC-150 MiniCycler	Peltier	No	100	100
Stratagene	RoboCycler 40	B	No	99	4
Stratagene	RoboCycler Gradient 40	B	C	99	4
Stuart Scientific	SPCR1 GENE-TECH	Peltier	No	48	29
Stuart Scientific	SPCR3 MINI-GENE	Peltier	No	48	29
Techne	Cyclogene	Peltier	No	99	99
Techne	Gene E	Air	No	99	9
Techne	PHC-1A	Water	No	63	3
Techne	PHC-2	Water	No	D	D
Techne	MW-1A	Water	No	63	3
Techne	MW-2	Water	No	D	D

B, One chilled block, three heated blocks, cooling between cycle segments is by robotic transfer between blocks; chilled block is cooled by Peltier effect.

C, By virtue of temperature gradient across one heated block.

D, Variable, limited by memory capacity.

n.i., no information.

Table 3. Thermal cyclers: III

Supplier/manufacturer	Product name	Number of repeat cycles per program	Program to program linking	Incubation incrementation/decrementation	Interchangeable reaction blocks
Applied Biosystems	GeneAmp PCR system 2400	1–99	Yes	Yes	No
Applied Biosystems	GeneAmp PCR system 9600	1–99	Yes	Yes	No
Applied Biosystems	DNA Thermal Cycler 480	1–99	Yes	Yes	No
Applied Biosystems	DNA Thermal Cycler	1–99	Yes	Yes	No
Applied Biosystems	GeneAmp *in situ* PCR 1000	1–99	Yes	Yes	No
Appligene	Crocodile II	1–99	n.i.	n.i.	Yes
Appligene	Crocodile III	1–99	n.i.	n.i.	Yes
Biometra	TRIO–Thermoblock	1–99	Yes	Yes	No
Biometra	UNO–Thermoblock	1–99	Yes	Yes	Yes
Biometra	Personal Cycler	1–99	No	Yes	Yes
BioTherm	Bio-Oven III	1–99	Yes	Yes	Yes (trays)
Camlab	Thermojet +	1–99	Yes	Yes	No
Eppendorf	Mastercycler 5330	1–999	Yes	Yes	No
Ericomp	SingleBlock System	1–999	Yes	Yes	No
Ericomp	TwinBlock System	1–999	Yes	Yes	No
Ericomp	DeltaCycler I (SingleBlock Peltier)	1–999	Yes	Yes	No
Ericomp	DeltaCycler II (TwinBlock Peltier)	1–999	Yes	Yes	No
Ericomp	PowerBlock System	1–999	Yes	Yes	No
Ericomp	SingleStar System	1–999	Yes	Yes	No

Company	Model			
Ericomp	TwinStar System	1–999	Yes	Yes	No
GL Applied Research	GTC-2	1–99	No	Yes	No
Grant	Autogene II	1–99	Yes	Yes, by program linking	Yes (racks)
Hybaid	Omnigene	1–99	Not necessary	Yes	No
Hybaid	Omnislide	1–99	Not necessary	Yes	No
Hybaid	Thermal reactor	1–99	Not necessary	No	No
Idaho Technology	1605 Air Thermo–Cycler	1–99	Yes	No	No
Integrated Separation Systems	Pro-Oven I	1–99	Not necessary	Yes	Yes (racks)
Integrated Separation Systems	Pro-Oven III	1–99	Not necessary	Yes	Yes (racks)
MJ Research	PTC-100	1–10 000	Not necessary	Yes	Yes
MJ Research	PTC-150 MiniCycler	1–10 000	Not necessary	Yes	Yes
Stratagene	RoboCycler 40	1–99	Yes	No	No
Stratagene	RoboCycler Gradient 40	1–99	Yes	No	No
Stuart Scientific	SPCR1 GENE–TECH	1–100	Yes	n.i.	Yes
Stuart Scientific	SPCR3 MINI–GENE	1–100	Yes	n.i.	Yes
Techne	Cyclogene	1–99	Yes	Yes, by program linking	Yes
Techne	Gene E	1–99	Yes	Yes	No
Techne	PHC-1A	1–99	No	Yes	No
Techne	PHC-2	D	Yes	No	No
Techne	MW-1A	1–99	No	Yes	No
Techne	MW-2	D	Yes	No	No

D, See footnote for *Table 2*. n.i., no information.

Instruments

Table 4. Thermal cyclers: IV

Supplier/manufacturer	Product name	Heated lid	Format
Applied Biosystems	GeneAmp PCR system 2400	Yes	E
Applied Biosystems	GeneAmp PCR system 9600	Yes	F
Applied Biosystems	DNA Thermal Cycler 480	No	G
Applied Biosystems	DNA Thermal Cycler	No	G
Applied Biosystems	GeneAmp *in situ* PCR 1000	N/A	H
Appligene	Crocodile II	Optional	I
Appligene	Crocodile III	Yes	I
Biometra	TRIO-Thermoblock	No	J
Biometra	UNO-Thermoblock	Optional	K
Biometra	Personal Cycler	Optional	L
BioTherm	Bio-Oven III	N/A	M
Camlab	Thermojet +	No	N
Eppendorf	Mastercycler 5330	Under development	O
Ericomp	SingleBlock System	Optional	P
Ericomp	TwinBlock System	Optional	Q
Ericomp	DeltaCycler I (SingleBlock Peltier)	Optional	R
Ericomp	DeltaCycler II (TwinBlock Peltier)	Optional	S
Ericomp	PowerBlock System	Optional	T
Ericomp	SingleStar System	No	U
Ericomp	TwinStar System	No	V
GL Applied Research	GTC-2	Optional	W
Grant	Autogene II	N/A	X
Hybaid	Omnigene	Optional	Y

Manufacturer	Model	Feature	Code
Hybaid	Omnislide	N/A	Z
Hybaid	Thermal reactor	No	AA
Idaho Technology	1605 Air Thermo-Cycler	N/A	AB
Integrated Separation Systems	Pro-Oven I	N/A	AC
Integrated Separation Systems	Pro-Oven III	N/A	AD
MJ Research	PTC-100	Yes	AE
MJ Research	PTC-150 MiniCycler	No	AF
Stratagene	RoboCycler 40	No	AG
Stratagene	RoboCycler Gradient 40	No	AG
Stuart Scientific	SPCR1 GENE-TECH	Under development	AH
Stuart Scientific	SPCR3 MINI-GENE	n.i.	AI
Techne	Cyclogene	Optional	AJ
Techne	Gene E	Optional	AK
Techne	PHC-1A	No	AA
Techne	PHC-2	No	AL
Techne	MW-1A	No	AM
Techne	MW-2	No	AM

All thermocyclers (with the exception of the Hybaid Thermal Reactor, the Idaho Technology 1605 Air Thermo-Cycler, the Techne PHC-1A and MW-1A) have a pause facility, useful for hot-start reactions if wax beads are not used (see Chapter 8, Section 2.2). All except the Eppendorf Mastercycler 5330, the Integrated Separation Systems Pro-Oven I, the Stratagene RoboCycler 40 and RoboCycler Gradient 40 have a printer or recorder output. The Camlab Thermojet+ can operate a slave unit via a smart card and single or twin units can be controlled by the Hybaid omnigene.

N/A, not applicable.

n.i., no information.

In the footnotes below, 'or' = alternative formats for an individual reaction block; 'or' = alternative formats available using alternative reaction blocks, trays or racks and 'and' = simultaneous processing of alternative formats.

Continued

Instruments

Table 4. Thermal cyclers: IV, *continued*

E, 24 × 0.2 ml MicroAmp tubes.
F, 96 × 0.2 ml MicroAmp tubes.
G, 48 × 0.5 ml tubes.
H, 10 × microscope slides (maximum three samples per slide).
I, 54 × 0.5 ml tubes, or 96-well microtiter plate or 96 × 0.2 ml tubes or four microscope slides using alternative reaction blocks.
J, 3 × 20 × 0.5 ml tubes.
K, 96 × 0.2 ml tubes or 96-well microtiter plate or 40 × 0.5 ml tubes or four microscope slides using alternative reaction blocks.
L, 20 × 0.5 ml tubes or 48 × 0.2 ml tubes using alternative reaction blocks.
M, 236 × 0.5 ml tubes or 132 × 2.0 ml tubes or 6 × 96-well microtiter plates or 30 microscope slides or glass capillaries (100s) using alternative reaction trays.
N, any combination of 25 × 0.5 ml tubes, 36 × 0.2 ml tubes or flatplate–depending on choice of units.
O, two blocks, each 27 × 0.5 ml tubes.
P, 60 × 0.5 ml tubes and 45 × 0.2 ml tubes or 96-well microtiter plate or 29 × 0.5 ml tubes and 6 × microscope slides using alternative reaction blocks.
Q, any combination of two formats in O.
R, 60 × 0.5 ml tubes or 96 × 0.2 ml tubes or 96-well microtiter plate using alternative reaction blocks.
S, any combination of two formats in Q.
T, 25 × 0.2 ml tubes or 16 × 0.5 ml tubes or 16 × 0.2 ml tubes or 8 × 0.5 ml tubes using alternative reaction blocks.
U, 60 × 0.2 ml tubes or 96 × 0.2 ml tubes or 96-well microtiter plate using alternative reaction blocks.
V, any combination of two formats in R.
W, 48 × 0.5 ml tubes or 48 × 0.5 ml tubes and 41 × 0.2 ml tubes or 96-well microtiter plate or four microscope slides, depending on reaction block.
X, 50 × 1.5 ml tubes or 50 × 0.5 ml tubes or 96-well microtiter plate or five microscope slides using alternative reaction racks.
Y, 48 × 0.5 ml tubes, 96-well microtiter plate or up to four microscope slides, depending on choice of master unit and slave units.
Z, 20 microscope slides.
AA, 54 × 0.5 ml tubes.

AB, 48 capillary tubes in microtiter format.
AC, 100 × 0.5 ml tubes *or* 2 × 96-well microtiter plates *or* 20 microscope slides using alternative reaction racks.
AD, 234 × 0.5 ml tubes *or* 4 × 96-well microtiter plates *or* 20 microscope slides using alternative reaction racks.
AE, 60 × 0.5 ml tubes *or* 96-well microtiter plate *or* 96 × 0.2 ml tubes *or* 12 microscope slides, depending on model.
AF, 16 × 0.5 ml tubes *or* 25 × 0.2 ml tubes depending on reaction block.
AG, 40 × 0.5 ml tubes.
AH, 39 × 0.5 ml tubes *or* 96-well microtiter plate, depending on reaction block.
AI, 15 × 0.5 ml tubes *or* 32-well 4 × 8 microtiter plate *or* 32 × 0.2 ml tubes *or* two microscope slides using alternative reaction blocks.
AJ, 40 × 0.5 ml tubes *or* 40 × 0.25 ml tubes *or* 40 × 0.2 ml tubes *or* 96 × 0.25 ml tubes *or* 96 × 0.2 ml tubes *or* 4 × microscope slides *or* 96-well microtiter plate, depending on reaction block.
AK, 40 × 0.5 ml tubes *or* 40 × 0.25 ml tubes *or* 40 × 0.2 ml tubes *or* 96 × 0.25 ml tubes *or* 96 × 0.2 ml tubes *or* 4 × microscope slides *or* 96-well microtiter plate, depending on model.
AL, 54 × 0.5 ml tubes *or* 24 × 1.5 ml tubes.
AM, 96-well microtiter plate.

Instruments

Chapter 4 NUCLEIC ACID SUBSTRATES – C.R. Newton

This chapter gives the PCR user the information required to assess the preparation and amount of template to use in a reaction. There are many published methods for DNA and RNA extraction. There are also many sources from which template nucleic acid can be obtained (see *Table 1* for examples). It therefore follows that in the space afforded here it is not practical to list and compare the different extraction methods for all template sources. Most modern methods of nucleic acid extraction are based on the affinity purification of nucleic acids, and in most instances this eliminates the need for organic extraction using hazardous chemicals such as phenol. It also reduces centrifugation steps and sample handling in general. This is important since it serves to reduce the potential of cross-contamination of samples.

For many PCR applications quite crude nucleic acid (EDTA) (lavender cap) blood collection tube (e.g. Sarstedt 49.355.001).
2. Take a 200 µl aliquot into a screw-capped microcentrifuge tube (e.g. Sarstedt 72.692).
3. Freeze the aliquot then thaw to lyse the erythrocytes.
4. Centrifuge in a microcentrifuge at full speed for 2 min and discard the supernatant.
5. Wash the cell pellet with 200 µl phosphate-buffered saline (PBS) and repeat step 4 (optional).
6. Add 200 µl water.
7. Place in boiling waterbath or 100°C hot-block for 5 min.
8. Centrifuge in a microcentrifuge at full speed for 5 min.
9. Use 10–30 µl of the supernatant per PCR.

1.2 Buccal wash

1. Centrifuge the whole sample at 10000 *g* for 10 min and

preparations are perfectly adequate, due to the fact that the sample may be diluted to reduce the concentration of any PCR inhibitors. Because of the amplification achieved by PCR, the dilution factor can usually be very large. The obvious exceptions are situations where the number of target DNA sequences may be initially low with respect to the starting sample volume (e.g. testing for pathogens in environmental samples or the monitoring of residual disease in patients undergoing cancer therapy).

A reasonably generic and simple DNA isolation procedure exists that provides DNA of sufficient quality for PCR in most applications. Some examples of how this may be adapted from one source to another are given below.

1 Examples of crude preparations for isolating DNA for PCR

1.1 Blood

1. Collect sample in ethylenediaminetetraacetic acid discard the supernatant.
2. Resuspend the cell pellet in 200 µl PBS and transfer to a screw-capped microcentrifuge tube (e.g. Sarstedt 72.692).
3. Continue from step 4 as described for blood.

1.3 Tissue culture cell suspensions

1. Take a 200 µl aliquot into a screw-capped microcentrifuge tube (e.g. Sarstedt 72.692).
2. Centrifuge in a microcentrifuge at full speed for 2 min and discard the supernatant.
3. Resuspend the cell pellet in 200 µl PBS.
4. Continue from step 4 as described for blood.

1.4 Bacterial liquid culture

1. Take a 200 µl aliquot into a screw-capped microcentrifuge tube (e.g. Sarstedt 72.692).
2. Centrifuge in a microcentrifuge at full speed for 2 min and aspirate off and discard the supernatant.

3. Add 200 µl water.
4. Continue from step 7 as described for blood, using 1–10 µl per PCR.

1.5 Bacterial colonies
1. Pick a colony into 200 µl water in a screw-capped microcentrifuge tube (e.g. Sarstedt 72.692).
2. Continue from step 7 as described for blood, using 1–10 µl per PCR.

RNA isolation procedures and quantitative RT-PCR methods are listed comprehensively in ref. 4.

2 Nucleic acid extraction kits

A large variety of kits exist for the extraction and purification of nucleic acids; these are outlined in *Table 2* (mRNA), *Table 3* (total RNA) and *Table 4* (DNA). The kits available for reverse transcriptase (RT) PCR cDNA synthesis are described in Chapter 14, *Table 16*.

be extrapolated roughly, taking into account the decrease in PCR efficiency after the amplification has passed the linear phase.

In addition, Clontech supplies purified total RNAs, poly(A)+ RNAs, cDNAs and cDNAs ligated to an oligonucleotide anchor ready for 5' rapid amplification of cDNA ends (RACE) (*Table 5*). These have been prepared from a variety of species and tissue sources. Suppliers of genomic DNAs include Bios Corporation (variety of human–rodent hybrid cell lines for chromosome localization studies, variety of species), Clontech (variety of species and transformed cell lines), Novagen (variety of species), Promega (variety of species), Sigma (variety of bacteria, *Saccharomyces cerevisiae* and Balb/C mouse).

Tables 6–8 show genome data for micro-organisms, plants and animals, respectively. The tables also show the calculated quantity of DNA required for a given number of genome equivalents for respective organisms.

Table 9 shows the extent of amplification for a range of PCR efficiencies. This table assumes one molecule of template in the PCR reaction. Initial target molecules per reaction and number of cycles required for a given efficiency can therefore

Table 1. DNA yields from different human tissue sources

Source of DNA	Amount typically used	Typical yield
Purified genomic DNA	10–500 ng	10–500 ng
Whole blood	30 µl	0.5–1 µg
Cell suspensions	5 × 10⁵ cells	2.0–5 µg
Buccal cells	One mouth wash	0.1–1 µg
Chorionic villus biopsy	Small frond (approx. 5 mg)	1–3 µg
Blood (Guthrie) spot	Half a 5 mm spot	0.5–1 µg
Semen	30 µl	5–10 µg
Hair roots	One root	10–200 ng
Tissue blocks	50 mg	0–10 µg

Table 2. mRNA isolation kits

Supplier	Extraction principle	Comments
Advanced Biotechnologies	Oligo-(dT)-cellulose spun column	No intermediate extraction from total RNA required, suitable for extraction from cells or tissue
Amersham	Capture on to oligo-(dT) membrane	Membrane added to cell lysate, removed and washed; comprises part of kit allowing subsequent RT-PCR
BCL	Oligo-(dT)-cellulose column	Extraction from total RNA
Clontech	Oligo-(dT)-cellulose spun column	Extraction from total RNA
5' to 3'	Oligo-(dT)-cellulose spun column	Extraction from total RNA
5' to 3'	Guanidinium isothiocyanate/ phenol extraction/ oligo-(dT)-cellulose spun column	No intermediate extraction from total RNA required, suitable for extraction from cells or tissue, organic extractions performed using 'phase lock gel' to physically separate organic and aqueous phases
5' to 3'	Oligo-(dT)-cellulose column	Extraction from total RNA
Genhunter	RNase free DNase I digestion	Purification of RNA contaminated with chromosomal DNA; important for successful DDRT-PCR
GIBCO-BRL	Guanidinium isothiocyanate/oligo-(dT)-cellulose spun column	No intermediate extraction from total RNA required, suitable for extraction from cells or tissue
Invitrogen	Detergent/protein degradation/ oligo-(dT)-cellulose spun column	No intermediate extraction from total RNA required, suitable for extraction from cells or tissue
Pharmacia	Guanidinium thiocyanate/oligo-(dT)-cellulose spun column	No intermediate extraction from total RNA required, suitable for extraction from cells or tissue
Promega	Guanidinium isothiocyanate/ biotinylated oligo-(dT)-cellulose/ streptavidin magnetic beads	No intermediate extraction from total RNA required, fast, efficient, suitable for extraction from cells or tissue

Supplier		Comments
Promega	Oligo-(dT)-cellulose/streptavidin magnetic beads	Extraction from total RNA
Qiagen	Latex particles bound with oligo-(dT)$_{30}$	Extraction from total RNA
Qiagen	Latex particles bound with oligo-(dT)$_{30}$	No intermediate extraction from total RNA required, suitable for extraction from cells or tissue
Sigma	Oligo-(dT)-cellulose spun column	No intermediate extraction from total RNA required, suitable for extraction from cells or tissue
Stratagene	Guanidinium isothiocyanate/oligo-(dT)-cellulose column	No intermediate extraction from total RNA required, suitable for extraction from cells or tissue

Table 3. Total RNA isolation kits

Supplier	Extraction principle	Comments
Clontech	Guanidinium thiocyanate	Recovery: 1–2 mg g^{-1} tissue
Clontech	AGPC	RNA extraction from plant tissue
Clontech	Guanidinium isothiocyanate/phenol chloroform	Minimal genomic DNA contamination; isolates RNA from very small amounts of starting material – microscale
5′ to 3′	Guanidinium isothiocyanate/phenol chloroform	Suitable for micro amounts of cells grown in 6-, 12-, 24-, 48- or 96-well plates
5′ to 3′	Guanidinium isothiocyanate/phenol chloroform	Suitable for small or large (depending on kit choice) quantities of tissue or cells, organic extractions performed using 'phase lock gel' to physically separate organic and aqueous phases
5′ to 3′	Guanidinium isothiocyanate/ centrifugation through CsCl cushion	

Continued

Table 3. Total RNA isolation kits, *continued*

Supplier	Extraction principle	Comments
GIBCO-BRL	Guanidinium thiocyanate/silica membrane spin carttridge	Suitable for small quantities of tissue or cells
Invitrogen	AGPC	Suitable for a wide variety of tissue sources and cells
Microprobe	Guanidinium thiocyanate/nuclease binding matrix	Co-purifies DNA, suitable for variety of cells, whole blood and Gram-negative bacteria
Pharmacia	Guanidinium thiocyanate/CsTFA	Suitable for a wide variety of tissue sources and cells, including plant material; particularly suitable for ribonuclease-rich samples
Promega	AGPC	Suitable for a wide variety of tissue sources and cells; particularly suitable for ribonuclease-rich samples
Sigma	Glass particle suspension	Suitable for recovery from agarose or solution
Stratagene	Guanidinium isothiocyanate/phenol chloroform	Suitable for a wide variety of tissue sources and cells
Stratagene	Guanidinium isothiocyanate/phenol chloroform	Suitable for micro amounts of tissues or cells

Abbreviations: AGPC, acid guanidinium thiocyanate–phenol–chloroform [1]; CsTFA, cesium trifluoroacetate.

Table 4. DNA isolation kits

Supplier	Extraction principle	Comments
BIO 101	Nuclei isolation	Specifically designed for plant genomic DNA isolation free of mitochondrial and chloroplast DNA
BioRad	Proprietry matrix that absorbs impurities	Suitable for direct isolation from blood, cultured cells and bacteria
Clontech	Glass particle suspension	Suitable for direct isolation from cells or tissues
5' to 3'	SDS/proteinase K/phenol chloroform	Organic extractions performed using 'phase lock gel' to physically separate organic and aqueous phases
Gentra Systems	Salting out proteins	Suitable for direct isolation from blood, body fluids, tissues, cultured cells, yeast, plant tissue and bacteria
GIBCO-BRL	Pre-PCR incubation with thermophilic protease	Suitable for small volumes of blood, cells or tissues
GIBCO-BRL	Detergent/SDS/proteinase K	Isolation from blood or cells
GIBCO-BRL	Silica membrane spin cartridge	ng to 50 µg per cartridge
Hoefer	Detergent	Suitable for small volumes of blood, cells or tissues
Invitrogen	Detergent	Suitable for small volumes of blood, cells or tissues
Microprobe	Guanidinium thiocyanate/nuclease binding matrix	Co-purifies RNA, suitable for variety of cells, whole blood and Gram-negative bacteria
Pharmacia	Guanidinium isothiocyanate/anion exchange spin column	Suitable for direct isolation from tissues, cultured cells, whole insects, plants and bacteria
Qiagen	Silica gel anion-exchange resin chromatography	Suitable for direct isolation from blood
Qiagen	Silica gel anion-exchange resin chromatography	Suitable for direct isolation from cultured cells
Qiagen	Silica gel anion-exchange resin spin column	Suitable for small volumes blood and cultured cells
Sigma	Detergent	Suitable for direct isolation from blood, cell cultures and tissues
Stratagene	Salting out proteins	Suitable for small volumes of blood
Stratagene	Salting out proteins	Suitable for direct isolation from blood, cultured cells and tissues

Table 5. cDNA kits

Supplier	1st strand synthesis	Comments
BCL Clontech	Oligo(dT), random hexamers or custom primer 5' RACE, custom primer	mRNA isolation from total RNA not required Anchor primer includes EcoRI site and sequence for directional cloning into pDIRECT vector (see Chapter 13, *Figure 11*)
5' to 3' GIBCO-BRL	Oligo(dT), random hexamers or custom primer Oligo(dT), random hexamers or custom primer 3' RACE, oligo(dT) primed	mRNA isolation from total RNA not required RNase H activity eliminated from reverse transcriptase (good yields of full-length cDNA) Product aliquot can go directly into PCR mRNA isolation from total RNA not required mRNA isolation from total RNA not required Universal amplification primer fused to uracil-containing sequence compatible with pAMP vectors (see Chapter 13, *Figures 12–15*)
Invitrogen Pharmacia	Oligo(dT), random hexamers or custom primer Oligo(dT), random hexamers or custom primer Oligo(dT) primed	Preliminary total RNA or mRNA isolation required 400 bp spun column size selection, includes EcoRI and NotI adaptors 300 bp spun column size selection, includes EcoRI and NotI adaptors
Promega	Random hexamer primed or Oligo(dT) primed or Oligo(dT) primed (XbaI adapted) or Oligo(dT) primed (NotI adapted)	Ligation of EcoRI adaptor allows directional cloning Ligation of EcoRI adaptor allows directional cloning

Table 6. Genomes and DNA substrates for PCR (micro-organisms)

Group	Organism	Genome size (kbp)[a]	% Single-copy DNA[a]	Mass DNA for 10^3 genome copies (ng)	No. copies of single-copy DNA ng^{-1}
Bacteria	*Haemophilus influenzae*	1200		1.29×10^{-3}	780×10^3
	Escherichia coli	4000		4.28×10^{-3}	230×10^3
	Bacillus megaterium	30 000		32×10^{-3}	30×10^3
Fungi	*Dictyostelium discoideum*	47 000	70	50×10^{-3}	20×10^3
	Saccharomyces carlsbergensis	15 000	89	16×10^{-3}	62×10^3
	Saccharomyces cerevisiae	20 000		21×10^{-3}	47×10^3
	Aspergillus nidulans	25 400	97	27×10^{-3}	37×10^3
	Hansenula holstii	10 300		11×10^{-3}	91×10^3
	Torulopsis holmii	21 600		23×10^{-3}	43×10^3
	Phycomyces blakesleeanus	30 100		32×10^{-3}	31×10^3
	Achlya bisexualis	41 400		44×10^{-3}	23×10^3
Protozoa	*Tetrahymena pyriformis*	190 000		203×10^{-3}	5×10^3
	Trypanosoma brucei	25 000	68	27×10^{-3}	37×10^3
	Trypanosoma cruzi	25 000	23	27×10^{-3}	37×10^3

[a]Data taken from *Molecular Biology Labfax*, see ref. 2, and papers cited therein.

Table 7. Genomes and DNA substrates for PCR (plants)

Group	Organism	Genome size (kbp)[a]	% Single-copy DNA[a]	Mass DNA for 10^3 genome copies (ng)	No. copies of single-copy DNA ng^{-1}
Angiospermae	Arabidopsis thaliana	190 000		0.2	4900
	Oryza sativa	565 000		0.6	1650
	Lycopersicon esculentum	700 000		0.8	1330
	Daucus carota ssp. carota	950 000		1.0	980
	Brassica napus	1 500 000		1.6	620
	Medicago sativa	1 600 000		1.7	580
	Solanum tuberosum	2 000 000		2.1	470
	Nicotiana tabacum	3 500 000		3.8	270
	Pisum sativum	4 700 000		5.0	200
	Secale cereale	8 275 000		8.9	110
	Zea mays	15 000 000		16.1	60
	Allium cepa	17 000 000		18.2	54
	Tulipa polychroma	23 000 000		24.6	40
	Lilium davidii	40 000 000		42.9	23
	Fritillaria assyriaca	120 000 000		128.5	8
	Lolium multiflorum		36		

[a] Data taken from *Molecular Biology Labfax*, see ref. 2, and papers cited therein.

Table 8. Genomes and DNA substrates for PCR (animals)

Group	Organism	Genome size (kbp)	% Single-copy DNA	Mass DNA for 10^3 genome copies (ng)	No. copies of single-copy DNA ng^{-1}
Arthropoda	Limulus polyphemus	2 650 000	70	2.8	360
	Plagusia depressa	1 400 000		1.5	670
	Drosophila melanogaster	165 000	60	0.2	5500
	Musca domestica	840 000		0.9	1110
	Bombyx mori	490 000	90	0.5	1900
	Locusta migratoria	5 000 000		5.4	190
Mollusca	Fissurella bandadensis	470 000		0.5	1990
	Tectorius muricatus	630 000		0.7	1480
	Aplysia californica	1 700 000		1.8	560
Echinodermata	Strongylocentrotus purpuratus	845 000	75	0.9	1100
Protochordata	Asidea atra	150 000		0.2	6220
Pisces	Rutilus rutilus	4 500 000	54	4.8	210
	Carcharias obscurus	2 650 000		2.8	350
Amphibia	Bufo bufo	6 600 000	20	7.1	140
	Xenopus laevis	2 900 000	75	3.1	320
	Ambystoma mexicanum	35 700 000		38.3	26
	Notophthalamus viridescens	42 500 000		45.5	22
Reptilia	Natrix natrix	2 350 000	47	2.5	400
	Python reticulatus	1 600 000	71	1.7	590
Aves	Gallus domesticus	1 125 000	80	1.2	830
Mammalia	Homo sapiens	2 800 000	64	3.0	330

Continued

Table 8. Genomes and DNA substrates for PCR (animals), *continued*

Group	Organism	Genome size (kbp)	% Single-copy DNA	Mass DNA for 10^3 genome copies (ng)	No. copies of single-copy DNA ng^{-1}
Mammalia	*Mus musculus*	3 300 000	70	3.5	290
	Microtus agrestis	3 000 000		3.2	310
	Peromyscus eremicus	4 420 000		4.7	210
	Dipodymys ordii monoensis	5 200 000		5.6	180

Table 9. Efficiencies

Cycle no.	PCR efficiency (%)									
	10	20	30	40	50	60	70	80	90	100
1	1.1	1.2	1.3	1.4	1.5	1.6	1.7	1.8	1.9	2.0
5	1.6	2.5	3.7	5.4	7.6	10.5	14.2	18.9	24.8	32.0
10	2.6	6.2	13.8	28.9	57.7	110.0	201.6	357.0	613.1	1.0×10^3
15	4.2	15.4	51.2	155.6	437.9	1.2×10^3	2.9×10^3	6.7×10^3	1.5×10^4	3.3×10^4
20	6.7	38.3	190.0	836.7	3.3×10^3	1.2×10^4	4.1×10^4	1.3×10^5	3.8×10^5	1.0×10^6
25	10.8	95.4	705.6	4.5×10^3	2.5×10^4	1.3×10^5	5.8×10^5	2.4×10^6	9.3×10^6	3.4×10^7
30	17.4	237.4	2.6×10^3	2.4×10^4	1.9×10^5	1.3×10^6	8.2×10^6	4.6×10^7	2.3×10^8	1.1×10^9
35	28.1	590.7	9.7×10^3	1.3×10^5	1.5×10^6	1.4×10^7	1.2×10^8	8.6×10^8	5.7×10^9	3.4×10^{10}
40	45.3	1.5×10^3	3.6×10^4	7.0×10^5	1.1×10^7	1.5×10^8	1.7×10^9	1.6×10^{10}	1.4×10^{11}	1.1×10^{12}
45	72.9	3.7×10^3	1.3×10^5	3.8×10^6	8.4×10^7	1.5×10^9	2.3×10^{10}	3.1×10^{11}	***3.5×10^{12}***	***3.5×10^{13}***
50	117.4	9.1×10^3	5.0×10^5	2.0×10^7	6.4×10^8	1.6×10^{10}	3.3×10^{11}	***5.8×10^{12}***	***8.7×10^{13}***	***1.1×10^{15}***

Figures shown in bold and italics are hypothetical, based on one molecule template in a 100 μl starting reaction mixture; PCR cannot proceed to generate this level of product since the reaction plateaus at 3–5 pmol [3] (3–5 pmol = 2–3 × 10^{12} molecules).

Chapter 5 THERMOSTABLE DNA POLYMERASES – C.R. Newton

DNA polymerases catalyze the synthesis of long polynucleotide chains from monomer deoxynucleoside triphosphates, using one of the original parental strands as a template for the synthesis of a new complementary strand. DNA synthesis proceeds in the 5' to 3' direction since the polymerization is from the 5' α-phosphate of the deoxynucleoside triphosphate to the 3' terminal hydroxyl group of the growing DNA strand. DNA polymerases require a short segment of DNA, or primer, to anneal to a complementary sequence to prime synthesis. Deoxynucleoside triphosphates (dNTPs) used in natural DNA synthesis normally comprise deoxyadenosine triphosphate (dATP), deoxycytidine triphosphate (dCTP), deoxyguanosine triphosphate (dGTP) and deoxythymidine triphosphate (dTTP). These dNTPs are covalently joined to the free hydroxyl group of the primer and form a newly synthesized strand complementary to the template. DNA polymerases may also have associated 3' to 5' and/or 5' to 3' exonuclease activities. For any given application these exonuclease activities should be considered. This is particularly the case with polymerases possessing a 3' to 5' activity, since this can cause degradation of the primers. However, the presence of this activity is associated with 'proofreading', therefore the fidelity of these enzymes is greater than that of polymerases without this activity.

(i) 5' to 3' DNA-dependent DNA polymerase, requiring a ssDNA template and a DNA or RNA primer with a 3'-OH terminus:

Thermostable DNA Polymerases

(ii) 5' to 3' exonuclease, degrading dsDNA or a DNA–RNA hybrid (including the RNA component) from a 5'-P terminus:

```
  5'    3'                          P 5' ---- 
P ———— ---                    →     ———————— +pN
  ———————                           ————————
  3'    5'                          3'      5'
```

(iii) 3' to 5' exonuclease, degrading ssDNA or dsDNA from a 3'-OH terminus:

Thermostable DNA polymerases can be added in a single addition at the beginning of the amplification process without further additions during the reaction. Commercially available thermostable polymerase enzymes are listed below in two categories, those without and those with

Prepared from Pyrococcus furiosus gene expressed in heterologous host.

Properties

1. Tenfold higher specific activity than *Pfu* (exo⁺);
2. No detectable 3' to 5' exonuclease proofreading activity;
3. Incorporates [α³⁵S]dATP 10 times more efficiently than does *Taq* DNA polymerase;
4. 95% active after 1 h incubation at 95°C.

Applications

1. High-temperature primer extension reactions;
2. [α³⁵S]dATP cycle sequencing (see Safety note);
3. Nucleotide analog incorporation.

Note – it is likely that deoxyuracil triphosphate (dUTP) alone instead of dTTP would be a poor substrate for this enzyme given that this is the case regarding *Pfu* (exo⁺) DNA polymerase (see below); it is therefore not recommended to use *Pfu* (exo⁻) DNA polymerase in conjunction with uracil DNA glycosylase (UDG) carry-over prevention.

associated 3′ to 5′ exonuclease activities. Sources and suppliers of thermostable DNA polymerases are given in *Table 1*, while *Table 2* lists their properties.

1 DNA polymerases without 3′ to 5′ exonuclease activity (3′ exo⁻)

In general, this group of enzymes adds a nontemplate-specified nucleotide to the 3′ termini of both strands of an amplicon (*Table 2*). This addition is predominantly an adenine (A) residue and can be exploited in TA cloning (see Chapter 13). The 3′ to 5′ exonuclease activity of DNA polymerases is associated with their ability to 'proofread' newly synthesized DNA. Hence the absence of the activity in this group of enzymes is coupled with increased misincorporation rates compared to the 3′ exo⁺ DNA polymerases described in section 2.

Pfu (exo⁻) DNA polymerase
Available from Stratagene (genetically modified recombinant form).

Psp(exo⁻) DNA polymerase
Available from New England Biolabs (Deep Vent® (exo⁻)) (genetically modified recombinant form).
Prepared from Pyrococcus sp. GB-D gene expressed in E. coli.

Properties

1. Has no 5′ to 3′ exonuclease activity;
2. Half-life of 23 h at 95°C;
3. Primer extension products up to 15 kb.

Applications

1. High-temperature primer extension reactions;
2. DNA cycle sequencing;
3. High-temperature dideoxy sequencing.

Taq DNA polymerase [1, 2]
Available from Amersham, Applied Biosystems (AmpliTaq®) [3], Boehringer Mannheim (recombinant form); Applied Biosystems, Appligene, GIBCO-BRL, Pharmacia, Promega, Stratagene (native form).

Thermostable DNA Polymerases

Prepared from *Thermus aquaticus* gene expressed in *E. coli* (recombinant form); *T. aquaticus* (native form).

Properties

1. Extension rate of 35–100 nucleotides sec^{-1} at 70–80°C;
2. Half-life of 40 min at 95°C;
3. High processivity;
4. Can incorporate deoxyuridine monophosphate (dUMP) for carry-over prevention
5. Has a 5' to 3' exonuclease activity.

Applications

1. High-temperature primer extension reactions;
2. Favored for sequencing PCR products by cycle sequencing;
3. PCR product cloning using TA vectors (see Chapter 13);
4. Amplification refractory mutation system (ARMS), ASA, PASA analyses.

N-terminal deletions of *Taq* DNA polymerase

Available from Applied Biosystems (AmpliTaq® Stoffel fragment) [4], Hoefer (Ultrotaq-BP®) (recombinant form).

Tbr DNA polymerase

Available from Amresco (Thermalase Tbr®), Flowgen (DynaZyme®), Hoefer (Hi-TAQ®), NBL Gene Sciences (native form).

Prepared from Thermus brockianus.

Properties

1. Half-life of 2.5 h at 96°C;
2. Twofold lower error frequency relative to *Taq* DNA polymerase;
3. High processivity;
4. Extension rate of 1–2 kb min^{-1} at 72°C;
5. Possesses reverse transcriptase activity (about 75% that of *Tth* DNA polymerase);
6. Reduced terminal transferase activity (after primer extension, a 15 min incubation at 72°C is required to add a 3' A to DNA products).

Applications

1. High-temperature primer extension reactions;
2. DNA sequencing;

3. Reverse transcription of RNA.

Tfl DNA polymerase

Available from Amersham (Hot *Tub*®), Promega (native form).

Prepared from Thermus flavus.

Properties

1. Primer extension products greater than 15 kb.

Applications

1. High-temperature primer extension reactions;
2. Cycle DNA sequencing.

Tli (exo⁻) DNA polymerase

Available from New England Biolabs (Vent® (exo⁻)) (genetically modified recombinant form).
Prepared from Thermococcus litoralis gene expressed in E. coli.

Properties

1. Deletion of around 289 amino acids from the N-terminus;
2. Half-life of 80 min at 95°C;
3. High processivity;
4. Has no intrinsic 5′ to 3′ exonuclease activity;
5. Optimal activity over a broader range of magnesium concentrations (2–10 mM) than unmodified *Taq* DNA polymerase, reducing the requirement for optimization of PCR conditions.

Applications

1. High-temperature primer extension reactions;
2. Used to improve amplification of templates known to be G–C rich, or with complex secondary structure;
3. Amplification of circular templates such as plasmid DNA;
4. Performing multiplex PCR;
5. Increases informity in random amplified polymorphic DNA (RAPD) analyses.

Thermostable DNA Polymerases

Properties

1. Has no intrinsic 5' to 3' exonuclease activity;
2. Half-life of 6.7 h at 95°C;
3. Primer extension products up to 15 kb.

Applications

1. High-temperature primer extension reactions;
2. High-temperature dideoxy DNA sequencing;
3. Thermal cycle sequencing.

Note – dUTP alone instead of dTTP is a poor substrate for this enzyme; it is therefore not recommended to use *Tli* (exo⁻) DNA polymerase in conjunction with UDG carry-over prevention [5].

Tth DNA polymerase

Available from Applied Biosystems (recombinant form); Amersham (TET-z®), Boehringer Mannheim, Novagen, Pharmacia, Promega (native form).

Prepared from Thermus thermophilus gene expressed in *E. coli* (recombinant form); *T. thermophilus* (native form).

Properties

4. Detection and analysis of gene expression at the RNA level;
5. Can incorporate dUMP for carry-over prevention;
6. High-temperature dideoxy DNA sequencing.

Notes

1. RT-PCR using this enzyme may be performed in a single-buffer or two-buffer system. The choice is dictated by the application. If the amplicon is to be cloned and/or for long cDNA synthesis, the two-buffer system should be used; that is, reverse transcription in the presence of Mn^{2+} followed by chelation of Mn^{2+} with EGTA and its replacement with Mg^{2+} for subsequent PCR. This ensures that each enzymatic activity is independently optimized. Where increased fidelity is not a major concern, a single buffer (see Chapter 16, *Table 14*, GeneAmp EZ r*Tth* RNA PCR Kit) system allows both activities to function without modification of the reaction mixture. Applications for this method are those not requiring high fidelity, such as detection and quantification of cellular or viral RNAs.

Properties

1. Reverse transcriptase activity in the presence of Mn^{2+} [6];
2. Half-life at 95°C is about 20 min;
3. 2 kb reverse transcripts;
4. Primer extension products up to 10 kb;
5. Both the polymerase and reverse transcriptase activities of this enzyme are reported to function in the presence of phenol [7]. The DNA polymerase activity remains functional in 2–5% (v/v) of phenol-saturated PBS buffer; the reverse transcriptase activity remains functional in up to 15% phenol-saturated PBS, enabling phenol-saturated aqueous phases of phenol partitions to be added directly to PCR reverse transcription reaction mixtures.

Applications

1. RNA (RT)-PCR (see note below);
2. Reverse transcription of RNA with high GC content and/or complex secondary structure;
3. High-temperature primer extension reactions;

2. r*Tth* DNA polymerase has been specially formulated with r*Tli* DNA polymerase (below) as a kit (see Chapter 14, GeneAmp XL PCR Kit) for 'long-PCR' capable of amplifying targets up to 40 kb.
3. r*Tth* DNA polymerase comprises the principal ingredient of the 'Single dA' tailing kit (Novagen) that facilitates the cloning of a DNA fragment with any type of ends into TA vectors (see Chapter 13).

2 DNA polymerases with 3′ to 5′ exonuclease activity (3′ exo$^+$)

DNA polymerases require the 3′ hydroxyl end of a base-paired polynucleotide strand (primer) on which to add further nucleotides. DNA molecules with a mismatched (i.e. not correctly base-paired) 3′ hydroxyl end can be corrected by the intrinsic 3′ to 5′ exonuclease activity of these enzymes, which will remove mismatched residues until a correctly base-paired terminus is generated. This is then an active template/primer substrate for the polymerase. A further consequence of the 3′ to 5′ exonuclease activity is degrada-

Thermostable DNA Polymerases

tion of primers from the 3′ end. For this reason either the enzyme or primers should be added last when setting up the reaction. This problem can be overcome by employing the 'hot-start' technique, or alternatively, by incorporating 3′-phosphorothioate linkages in the primers during synthesis, using standard oligonucleotide synthesis chemistry [8].

Pfu DNA polymerase [9]

Available from Stratagene (recombinant form); Stratagene (native form).
Prepared from P. furiosus gene expressed in heterologous host (recombinant form); *P. furiosus* (native form).

Properties

1. Intrinsic 5′ to 3′ exonuclease activity;
2. Half-life of 13 h at 95°C.

Applications

1. High-fidelity, high-temperature primer extension reactions;
2. Additive for long PCR;
3. Polishing PCR product 3′ additions [10];
4. DNA cycle sequencing;
5. High-temperature dideoxy sequencing;
6. Blunt end cloning.

Pwo DNA polymerase

Available from Boehringer Mannheim (native form).
Prepared from Pyrococcus woesei.

Properties

1. Has no 5′ to 3′ exonuclease activity;
2. Half-life of 2 h at 100°C.

Applications

1. High-fidelity, high-temperature primer extension reactions;
2. DNA cycle sequencing;
3. High-temperature dideoxy sequencing;
4. Blunt-end cloning.

Note – dUTP alone instead of dTTP is a poor substrate for this enzyme; it is therefore not recommended to use *Pwo*

4. *Not recommended for* high-temperature primer extension reactions employing primers containing deoxyinosine [11].

Note – dUTP alone instead of dTTP is a poor substrate for this enzyme; it is therefore not recommended to use *Pfu* (exo⁺) DNA polymerase in conjunction with UDG carry-over prevention [5].

Psp *DNA polymerase*

Available from New England Biolabs (Deep Vent®) (genetically modified recombinant form).

Prepared from Pyrococcus sp. *GB-D* gene expressed in *E. coli*.

Properties

1. Has no 5' to 3' exonuclease activity;
2. Half-life of 23 h at 95°C;
3. Primer extension products up to 8.2 kb.

Applications

1. High-fidelity, high-temperature primer extension reactions;

DNA polymerase in conjunction with UDG carry-over prevention.

Tli *DNA polymerase*

Available from New England Biolabs (Vent®) (recombinant form); Promega (native form).

Prepared from T. litoralis gene expressed in *E. coli* (recombinant form); *T. litoralis* (native form).

Properties

1. Has no intrinsic 5' to 3' exonuclease activity;
2. Half-life of 6.7 h at 95°C;
3. Primer extension products up to 13.2 kb.

Applications

1. High-fidelity, high-temperature primer extension reactions;
2. Blunt-end cloning;
3. As companion enzyme with r*Tth* DNA polymerase (above), r*Tli* DNA polymerase has been specially

Thermostable DNA Polymerases

formulated in a kit for 'long-PCR' (see Chapter 14, Section 7, GeneAmp XL PCR Kit).

Note – dUTP alone instead of dTTP is a poor substrate for this enzyme; it is therefore not recommended to use *Tli* DNA polymerase in conjunction with UDG carry-over prevention [5].

Tma DNA polymerase

Available from Applied Biosystems (UlTma®) (recombinant form).

Prepared from Thermotoga maritima gene expressed in E. coli.

Properties

1. No associated 5' to 3' exonuclease activity;
2. Half-life at 97.5°C is about 1 h;
3. Blunt-ended primer extension products.

Applications

1. High-fidelity, high-temperature primer extension reactions;
2. Blunt-end cloning.

Table 1. Thermostable DNA polymerases

DNA polymerase (generic name)	DNA polymerase (trade name)	Natural/recombinant	Source	Supplier
Pfu (exo$^+$ and exo$^-$)		Natural	P. furiosus	STG
Psp (exo$^+$ and exo$^-$)	Deep Vent (exo$^+$ and exo$^-$, NEB)	Recombinant	Pyrococcus sp. GB-D	NEB
Pwo		Natural	P. woesei	BM
Taq		Natural	T. aquaticus	ABI, APG, CLO, GIBCO-BRL, PHA, PRO, STG
Taq	AmpliTaq (ABI)	Recombinant	T. aquaticus	ABI, AME, BM
Taq preparations (N-terminal deleted)	AmpliTaq, Stoffel fragment (ABI), Ultrotaq-BP (HOE)	Recombinant	T. aquaticus	ABI, HOE
Tbr	Thermalase Tbr (AMR), DynaZyme (FLO), Hi-TAQ (HOE)	Natural	T. brockianus	AMR, FLO, HOE, NBL
Tfl	Hot Tub (AME)	Natural	T. flavus	AME, PRO
Tli		Natural	T. litoralis	PRO
Tli (exo$^+$ and exo$^-$)	Vent (exo$^+$ and exo$^-$, NEB)	Recombinant	T. litoralis	NEB
Tma	UlTma (ABI)	Recombinant	T. maritima	ABI
Tth		Recombinant	T. thermophilus	ABI
Tth	TET-z (AME)	Natural	T. thermophilus	ADB, AME, BM, CLO, PHA, PRO

Abbreviations: ABI, Applied Biosystems; ADB, Advanced Biotechnologies; AME, Amersham; AMR, Amresco; APG, Appligene; BM, Boehringer Mannheim; CLO, Clontech; FLO, Flowgen; HOE, Hoefer; NBL, NBL Gene Sciences; NEB, New England Biolabs; PHA, Pharmacia; PRO, Promega; STG, Stratagene.

Thermostable DNA Polymerases

Table 2. Properties of thermostable DNA polymerases

DNA polymerase	5' to 3' exonuclease activity	3' to 5' exonuclease activity	Reverse transcriptase activity	Resulting ends of amplicons
Pfu	No	Yes	n.i.	Blunt
Pfu (exo⁻)	No	No	n.i.	Blunt
Psp	No	Yes	n.i.	>95% blunt
Psp (exo⁻)	No	No	n.i.	70% blunt
Pwo	No	Yes	n.i.	Blunt
Taq	Yes	No	Weak	3'A
Taq (N-terminal deleted)	No	No	Weak	3'A
Tbr	Yes	No	Yes[a]	[b]
Tfl	n.i.	No	n.i.	n.i.
Tli	No	Yes	n.i.	>95% blunt
Tli (exo⁻)	No	No	n.i.	70% blunt
Tma	No	Yes	No	Blunt
Tth	Yes	No	Yes	3'A

n.i., no information.
[a] About 75% that of Tth DNA polymerase.
[b] After primer extension, a 15 min incubation at 72°C is required to add 3' A to DNA products.

Chapter 6 **PRIMERS** – C.R. Newton

1 Primer design

The aims of good primer design are to maximize both the specificity and efficiency of the amplification reaction. For some PCR applications there may be a trade off of one over the other, for example highly efficient amplifications are not needed in many diagnostic procedures, however specificity is very important in order to avoid misdiagnosis. Specificity and efficiency are determined by the following primer-dependent parameters:

1. Primer positions with respect to target (defining size of amplicon);
2. Primer positions with respect to target coding or non-coding sequence;
3. Primer positions with respect to conserved regions of coding sequence of a gene family;
4. Base composition of primers;
5. Primer length;
6. Melting temperature (T_m) of primers (a function of primer lengths and their base composition);
7. The degree of degeneracy within the primer(s);
8. The incorporation of nucleotide analogs (e.g. phosphorothioate-linked residues);
9. The incorporation of 'unusual' nucleotides (e.g. dUMP, deoxyinosine monophosphate (dIMP), universal nucleotide) [1], see Chapter 7, Section 2.6;
10. The need to add extra, noncomplementary sequence;
11. Mismatches within the primers;
12. Mismatches at the ends of primers;
13. Complementarity between the ends of primers (*inter* and *intra*).

Table 1 summarizes the key variables that require consideration for a given PCR application.

Computer programs can help in primer design (e.g. OLIGO, Primer Detective, Primer Select). These calculate the DNA duplex stability based on pairwise interacting nearest neighbor analysis. These programs also take into account other features such as regions of DNA complementarity and secondary structure, which are important to avoid when designing amplimers manually.

Primers used in PCR are generally between 18 and 30 nucleotides in length, which allows a reasonably high annealing temperature to be used. Exceptions are shorter for DDRT-PCR [2–3] and RAPD [4]; longer when additional 5' sequence has to be added such as restriction endonuclease recognition sites [5], GC clamps for denaturing gradient gel electrophoresis (DGGE) analyses [6] or promoter sequences for gene amplification with transcript sequencing (GAWTS) [7], RNA amplification with transcript sequencing (RAWTS) [8] and expression PCR (E-PCR) [9]. near the center of the amplimer. The other notable exception is the amplification refractory mutation system (ARMS) [10] where one ARMS primer is specifically mismatched with one of two alternative templates and matched with the other. In cases where degenerate primers have to be used the degeneracy should be kept to an absolute minimum at the 3' end. To achieve this, consider Chapter 14, *Tables 9* and *12*; see also Chapter 7 (modified nucleosides) for universal nucleoside alternatives.

2 Calculating extinction coefficients for oligonucleotides and converting OD readings to micrograms and nanomoles

(adapted courtesy of OSWEL custom oligonucleotide synthesis)

2.1 Molecular weight of DNA

Mol. wt = $(249 \times nA) + (240 \times nT) + (265 \times nG) + (225 \times nC) + (63 \times n-1) + 2$

where: nA is the number of adenine bases in the DNA sequence, etc., and n is the total number of bases. $(63 \times n-1)$ accounts for the molecular weight of the phosphate groups.

Example
The 20mer d(AGCTCTGAACGTAGCTCTGA):

mol. wt = $(5 \times 249) + (5 \times 240) + (5 \times 265) + (5 \times 225) + (63 \times 19) + 2$.

Molecular weight = 6094.

2.2 Calculation of the micromolar extinction coefficient, ε, at 264 nm

Where $\varepsilon = OD_{264}$ units in a solution of 1 µmole dissolved in 1 liter, that is:

$\varepsilon = OD_{264}$ units of 1 µmole

$\varepsilon = \{(8.8 \times nT) + (7.3 \times nC) + (11.7 \times nG) + (15.4 \times nA)\} \times 0.9$.

The determination of ε can be simplified by using *Table 2*.

Specificity is not increased by designing primers longer than 30 nucleotides. Indeed, primers in the range 18–24 nucleotides are extremely specific provided that the annealing temperature of the PCR is a few degrees below the actual T_m (as opposed to the calculated T_m, see below). The primers should, if possible, be made such that the GC content is in the range 40–60%. In addition, regions of unusual sequence, such as stretches of polypurines, polypyrimidines, repetitive motifs or significant secondary structure, should be avoided. Primer pairs should also be designed so that there is no complementarity of the 3' ends either *inter* or *intra* individual primers. This precaution will reduce the incidence of 'primer–dimer' formation. When designing primers manually, one method that assists in avoiding complementarity is, if possible, to pick potential amplimers that are deficient in one or different (but not complementary) bases, while maintaining the 40–60% GC content rule.

For some applications the primers may not be exactly complementary to the template. For example, in site-directed PCR mutagenesis, the mutated base(s) should be positioned

Example
The 20mer d(AGCTCTGAACGTAGCTCTGA):

$\varepsilon = \{(8.8 \times 5) + (7.3 \times 5) + (11.7 \times 5) + (15.4 \times 5)\} \times 0.9$.

Micromolar extinction coefficient, $\varepsilon = 194.4$

Note that it is necessary to multiply the extinction coefficient of the sum of the individual bases by 0.9. This is because the base stacking interactions in the single strand suppress the absorbance of DNA. This suppression is even greater in a duplex, and the multiplication factor for a self-complementary sequence is 0.8. *Table 2* accommodates for these factors. These correction factors are estimates for typical DNA sequences.

2.3 To convert OD units to milligrams

From Sections 2.1 and 2.2 we know that for the 20mer d(AGCTCTGAACGTAGCTCTGA), 6.09 milligrams (1 µmole) of the lyophilized solid will have an absorbance of 194.4 OD units at 264 nm.

$33 \times$ OD value = number of µg, **B**
B/A = number of µmoles
B/A $\times$ 1000 = number of nmoles.

For the sake of convenience an aliquot of the primer stock may be appropriately diluted so that a common volume of each primer solution is always added to every PCR reaction.

3 Calculating T_ms of oligonucleotides

Three common formulae are used for this purpose:

1. $T_m = [(A+T) \times 2°C + (G+C) \times 4°C]$.

   This was originally derived in a 1 M salt concentration for oligonucleotide hybridization assays [11]. Accurate for primers up to 20 nt. Many laboratories use annealing temperatures of 3–5°C below the T_m calculated, using this formula as a starting point for optimization experiments.

2. $T_m = 81.5 + 16.6(\log_{10}[J^+]) + 0.41(\%G+C)$
   $- (600/l) - 0.63(\%FA)$

   where $[J^+]$ = concentration of monovalent cations,

Therefore 6.09 mg = 194.4 OD units; therefore 1 mg = 194.4/6.09 = 32 OD units; and therefore 1 OD unit = 1/32 mg = 31.3 µg.

2.4 Simplified general calculations

Molecular weight of an oligonucleotide = length × 330 (**A**), therefore:
for a 20mer, molecular weight = 6600.

1 OD unit per 33 µg is an average figure for absorbance (**B**), therefore:
5 OD units = 165 µg.
From **B**, the number of micrograms = 33 × absorbance at 264 nm.

The number of micromoles = **B**/**A**, i.e.
165/6600 = 0.025, so 5 OD units of a 20mer = 0.025 µmoles,
5 OD units of a 20mer = 25 nanomoles = 165 µg.

To summarize
330 × oligo length = molecular weight, **A**

l = oligonucleotide length, FA = formamide [12]. This formula is suitable for oligonucleotides of 14–70 residues.

3. $T_p = 22 + 1.46(l_n)$

where T_p = optimized annealing temperature ±2–5°C, l_n = effective length of primer = 2(number of G or C) + (number of A or T) [13]. This formula is suitable for oligonucleotides of 20–35 residues.

Note

The calculated T_ms of amplimers are only approximately related to optimal annealing temperatures. Calculated T_ms therefore act as a reference point to begin experimentation. There may actually be up to a 12°C difference between the calculated T_m and the actual T_m. Ideal annealing temperatures are therefore determined empirically.

4 Primer labeling

Primers can be labeled at their 3′ termini by virtue of template annealing and primer extension using a DNA polymerase (see Chapter 11). In this situation 3′-end labeling is achieved using a dye- or radiolabeled dNTP or dideoxynucleoside triphosphate (ddNTP). Since the 3′ end of an amplimer is required for primer extension during PCR, enzymatically added labels are useful only at the 5′ end of an amplimer. This confines choice to the use of T4 polynucleotide kinase and a radioisotope labeled ATP or analog of ATP. Such compounds are shown in Chapter 14, *Table 4*. For primer synthesis and chemical labeling of amplimers, see Chapter 7.

4.1 Suppliers of amplimer panels

Advanced Biotechnologies, Applied Biosystems, Clontech, Genosys Biotechnologies, Novagen, Oncogene Science Inc., Operon Technologies Inc., R&D Systems and Research Genetics.

For amplimer custom synthesis and suppliers of amplimer synthesis reagents, see Chapter 7.

Table 1. Primer design

Technique	Important primer characteristics[a]
PCR using enzyme with 3' to 5' exonuclease activity	8
Contamination avoidance	9
Multiplex PCR	2, 6, 13
Restriction site incorporation	11
SOE (recombinant PCR)	10, 11, 13
RT-PCR	2, 7, 9
DDRT-PCR (mRNA differential display)	5
Mutagenesis	11
Expression PCR	10
GAWTS	10
RAWTS	10
Pathogen detection	1, 3, 6, 7
Diagnosing residual disease after cancer therapy	1, 3, 6
ARMS (ASA, PASA, ASP)	1, 3, 6, 12
Multiplex ARMS	1, 2, 3, 6, 12, 13
COP	5, 6, 11
Introduction of restriction sites	11, 12
Whole genome amplification	4, 5
Preimplantation diagnosis	1, 3, 6
DGGE	2, 10
RAPD (AP-PCR)	5

[a] As defined in Section 1.

Table 1. Micromolar extinction coefficient multiples

Number of residues	Nucleotide											
	T			A			G			C		
	ε	ε × 0.9	ε × 0.8	ε	ε × 0.9	ε × 0.8	ε	ε × 0.9	ε × 0.8	ε	ε × 0.9	ε × 0.8
1	8.8	7.9	7.0	15.4	13.9	12.3	11.7	10.5	9.4	7.3	6.6	5.8
2	17.6	15.8	14.1	30.8	27.7	24.6	23.4	21.1	18.7	14.6	13.1	11.7
3	26.4	23.8	21.1	46.2	41.6	37.0	35.1	31.6	28.1	21.9	19.7	17.5
4	35.2	31.7	28.2	61.6	55.4	49.3	46.8	42.1	37.4	29.2	26.3	23.4
5	44.0	39.6	35.2	77.0	69.3	61.6	58.5	52.7	46.8	36.5	32.9	29.2
6	52.8	47.5	42.2	92.4	83.2	73.9	70.2	63.2	56.2	43.8	39.4	35.0
7	61.6	55.4	49.3	107.8	97.0	86.2	81.9	73.7	65.5	51.1	46.0	40.9
8	70.4	63.4	56.3	123.2	110.9	98.6	93.6	84.2	74.9	58.4	52.6	46.7
9	79.2	71.3	63.4	138.6	124.7	110.9	105.3	94.8	84.2	65.7	59.1	52.6
10	88.0	79.2	70.4	154.0	138.6	123.2	117.0	105.3	93.6	73.0	65.7	58.4
11	96.8	87.1	77.4	169.4	152.5	135.5	128.7	115.8	103.0	80.3	72.3	64.2
12	105.6	95.0	84.5	184.8	166.3	147.8	140.4	126.4	112.3	87.6	78.8	70.1
13	114.4	103.0	91.5	200.2	180.2	160.2	152.1	136.9	121.7	94.9	85.4	75.9
14	123.2	110.9	98.6	215.6	194.0	172.5	163.8	147.4	131.0	102.2	92.0	81.8
15	132.0	118.8	105.6	231.0	207.9	184.8	175.5	158.0	140.4	109.5	98.6	87.6
16	140.8	126.7	112.6	246.4	221.8	197.1	187.2	168.5	149.8	116.8	105.1	93.4
17	149.6	134.6	119.7	261.8	235.6	209.4	198.9	179.0	159.1	124.1	111.7	99.3
18	158.4	142.6	126.7	277.2	249.5	221.8	210.6	189.5	168.5	131.4	118.3	105.1
19	167.2	150.5	133.8	292.6	263.3	234.1	222.3	200.1	177.8	138.7	124.8	111.0
20	176.0	158.4	140.8	308.0	277.2	246.4	234.0	210.6	187.2	146.0	131.4	116.8

Chapter 7 THE SYNTHESIS OF CHEMICALLY LABELED PCR PRIMERS
— T. Brown and D.J.S. Brown

Chemically labeled PCR primers are becoming increasingly important in experiments involving the capture of PCR products and subsequent nonradioactive detection (see Chapter 11). Here, we outline the chemical structures and means of attachment of a number of chemical groups commonly used to label synthetic oligonucleotides.

1 Oligonucleotide synthesis and purification

Automated solid-phase oligonucleotide synthesis is now a routine undertaking [1] and a wide variety of phosphoramidite monomers and active esters are available for the attachment of chemical labels to synthetic oligonucleotides. However, the monomers used in the chemical labeling of oligonucleotides are generally expensive and it is usually cost-effective to label several oligonucleotides simultaneously in order to minimize unit costs. Many of the monomers listed here are much less stable in solution than the standard A, G, C and T phosphoramidites and a certain degree of care is required in their use. The labeling of amino-functionalized oligonucleotides with active esters of fluorescent dyes and digoxigenin can be particularly tricky. The efficiency of labeling varies according to the nature of the DNA sequence and the length of the oligonucleotide. It is important to purify chemically labeled oligonucleotides by high pressure liquid chromatography (HPLC) or FPLC in order to remove failure sequences which, although unlabeled, are likely to retain biological activity [2].

(a)

(b)

(c)

Because of the above considerations, it is advisable to consider purchasing chemically modified PCR primers from a reputable commercial oligonucleotide synthesis service. In this way synthesis quality and product purity can be guaranteed. In addition, the real cost of purchasing labeled oligonucleotides may, in many cases, be less than the cost of purchasing the chemical labels and adding them to oligonucleotides 'in house'.

2 Chemical labels and labeled PCR primers

This chapter considers only chemical labels added to synthetic oligonucleotides by chemical methods (for enzymatic labeling methods see Chapter 6). It is, however, important to bear in mind that chemical labels can be added to unlabeled synthetic oligonucleotides by means of chemically labeled nucleoside triphosphates and enzymes such as DNA polymerase or terminal transferase. Enzymatic methods are useful for labeling large DNA fragments, but their utility is less clear in the case of short synthetic oligonucleotides.

Enzymatic labeling is normally carried out on a very small scale and tends to produce mixtures of various labeled and unlabeled products. This can give rise to problems in diagnostic applications when quantification and reproducibility are important factors.

Figures 1–7 show the chemical structures of a variety of chemical labeling groups. The list is far from exhaustive and contains only those groups commonly used at the time of writing.

2.1 Biotin monomers

The single addition biotin monomer [3] is widely used in applications where it is necessary to capture or detect biotinylated oligonucleotides by means of avidin or streptavidin–enzyme conjugates (*Figure 1a*). The multi-addition biotin monomer (*Figure 1b*) can be used to add several biotin moieties to the 5′ end of an oligonucleotide (*Figure 1c*) and this may be useful when detecting the biotin by means of an anti-biotin antibody–enzyme conjugate. However, it is not clear that multiple biotin groups offer a

Figure 1. (a) Single addition biotin phosphoramidite monomer. (b) Multiple addition biotin phosphoramidite monomer. (c) Structure of an oligonucleotide with three biotin molecules attached at the 5′ end.

significant advantage in avidin-based detection. The single addition biotin monomer is widely available and the multi-addition biotin monomer can be obtained from Clontech.

2.2 Digoxigenin

Digoxigenin has gained popularity as a labeling group for synthetic oligonucleotides [4–6]. An antibody–enzyme conjugate is available from Boehringer Mannheim to detect digoxigenin-labeled oligonucleotides. Chemical addition of digoxigenin is carried out after oligonucleotide synthesis by reaction of an aminohexyl-functionalized oligonucleotide (*Figure 2b*) with digoxigenin active ester (*Figure 2a*). The structure of the digoxigenin-labeled oligonucleotide is shown in *Figure 2c*. This type of post-synthetic labeling must be carried out with great care and HPLC purification of digoxigenin-labeled oligonucleotides is essential. This method only permits the addition of a single digoxigenin moiety to the 5′ end of an

Figure 2. Addition of the hapten digoxigenin to a synthetic oligonucleotide. This is carried out by reaction of the digoxigenin active ester to an aminohexyl-functionalized synthetic oligonucleotide. (a) Digoxigenin active ester. (b) Aminohexyl phosphoramidite and oligonucleotide: (i) monomer for the attachment of a primary aminohexyl group to the 5' end of a synthetic oligonucleotide; (ii) 5'-aminohexyl-functionalized oligonucleotide for the attachment of digoxigenin active ester to a synthetic oligonucleotide. (c) Structure of a synthetic oligonucleotide with digoxigenin attached at the 5' end.

Figure 3. (a) DNP phosphoramidite monomer for the single or multiple addition of dinitrophenyl haptens to the 5' end of synthetic oligonucleotides. (b) Single DNP group added to the 5' end of a synthetic oligonucleotide. (c) Three DNP groups added to the 5' end of a synthetic oligonucleotide.

oligonucleotide. There is no commercially available digoxigenin phosphoramidite monomer due to the chemical complexity of the label and its instability under the conditions of routine oligonucleotide synthesis. The digoxigenin active ester is supplied by Boehringer Mannheim and digoxigenin-labeled oligonucleotides can be obtained from suppliers who are licensed by Boehringer Mannheim.

2.3 Dinitrophenyl (DNP)

The DNP group is highly immunogenic and antibodies are available from a variety of sources. It is a useful hapten, which can easily be added to synthetic oligonucleotides (*Figure 3a*) by means of a commercially available multi-addition phosphoramidite monomer [7]. The structures of synthetic oligonucleotides bearing single or triple DNP labels are shown in *Figure 3b* and *c*. DNP-labeled oligonucleotides are bright yellow in color and this offers certain advantages during synthesis and purification. Anti-DNP antibody–enzyme conjugates are available from DakoPatts. Unconjugated rat monoclonal anti-DNP and enzyme-conjugated anti-rat antibodies are available from Novagen. The phosphoramidite monomer is available from Cruachem.

2.4 Fluorescent dyes

There are many methods by which fluorescein can be added to synthetic oligonucleotides [8–12]. In general, fluorescent

Chemically Labeled Primers

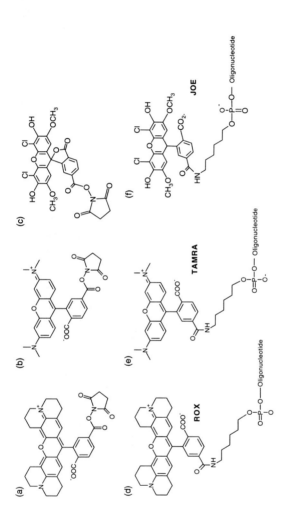

Figure 4. Applied Biosystems fluorescent dyes. (a)–(c) ABI fluorescent dye active esters that are attached to a synthetic oligonucleotide which has been pre-functionalized with a 5′-aminohexyl group: (a) Applied Biosystems ROX active ester; (b) Applied Biosystems TAMRA active ester; (c) Applied Biosystems JOE active ester. (d)–(f) Applied Biosystems fluorescent dyes attached to the 5′ end of synthetic oligonucleotides via a dye active ester and an amino-link phosphoramidite.

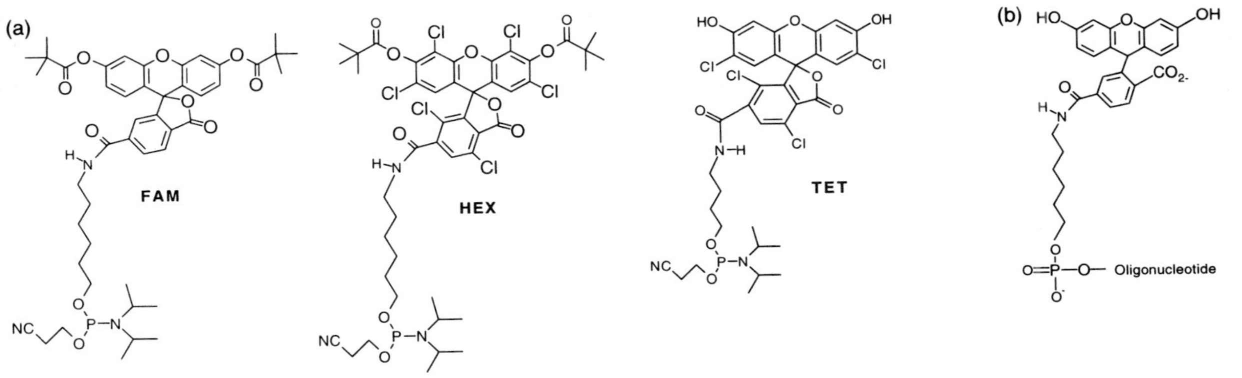

Figure 5. Applied Biosystems fluorescent dye phosphoriamidites – chlorinated derivatives of fluorescein. (a) FAM, HEX and TET; (b) FAM oligonucleotide.

dyes can be added post-synthetically to amino-functionalized oligonucleotides by means of active esters (*Figure 4a–c*) or, more conveniently, by the use of phosphoramidite monomers (*Figure 5a*) during solid-phase synthesis. In both cases the eventual linkage between the oligonucleotide and the dye is the same (*Figures 4d–f* and *5b*). A wide variety of fluorescent dyes are available for use in fluorescence microscopy or fluorescence spectroscopy.

Chemically Labeled Primers

post-synthetically. They will eventually be superseded by the more stable fluorescein-based phosphoramidites shown in *Figure 5*. *Table 1* shows the properties of this group of dyes.

Fluorescein-labeled oligonucleotides can be detected by anti-fluorescein antibody–enzyme conjugates and captured with immobilized anti-fluorescein antibodies. A multi-addition fluorescein monomer is available from Clontech (see Chapter 10) and this may be of use in antibody-based technologies. However, addition of multiple fluorescein moieties to the 5′ end of an oligonucleotide is not beneficial if fluorescent detection is to be employed, as fluorescence quenching occurs.

2.5 Other 5′ chemical labels

Single or multiple tetraethylene glycol or hexaethylene glycol phosphoramidites (*Figure 6a*) can be used to provide a spacer between hydrophobic groups and synthetic oligonucleotides (*Figure 6b*). This application is discussed in Chapter 10.

Reactive groups such as aminohexyl (*Figure 2b*) or thiohexyl

Figure 6. (a) Tetraethylene glycol phosphoramidite monomer which can be used as a hydrophilic spacer between a synthetic oligonucleotide and a reporter group. (b) The structure of a synthetic oligonucleotide attached to a reporter group (R) via a hydrophilic tetraethylene glycol spacer. Several tetraethylene glycol spacers can be added if necessary. (c) Thiol phosphoramidite monomer for the attachment of a thiohexyl reactive group to the 5′ end of a synthetic oligonucleotide. (d) A synthetic oligonucleotide with a thiol (thiohexyl) group attached to the 5′ end.

A very important application is in the field of DNA sequencing and genetic linkage analysis, where the dyes are excited by laser radiation and the labeled oligonucleotides can be detected with a very high degree of sensitivity. In the Applied Biosystems DNA sequencer several different families of labeled DNA fragments can be separated in the same lane of an electrophoresis gel and visualized by the characteristic absorption–emission frequencies of their dye labels. Each of the Applied Biosystems dyes in *Figures 4* and *5* has a characteristic emission frequency. ROX and TAMRA are rhodamine-based dyes and JOE is a derivative of fluorescein (*Figure 4a–c*). These dyes are unstable to the conditions of oligonucleotide synthesis and so must be added

(*Figure 6c*) can be added to the 5′ end of PCR primers, and these groups can be used to attach other reporter groups to oligonucleotides. This application has been illustrated previously for digoxigenin and fluorescent dyes (*Figures 2* and *4*).

A 5′-phosphate group can be added to any oligonucleotide using the monomer shown in *Figure 7*. If this monomer is added to a normal A, G, C or T synthesis column and the desired oligonucleotide is then synthesized, it has the effect of adding a 3′-phosphate to the sequence. This is of no practical value in PCR primers but may be useful in other applications.

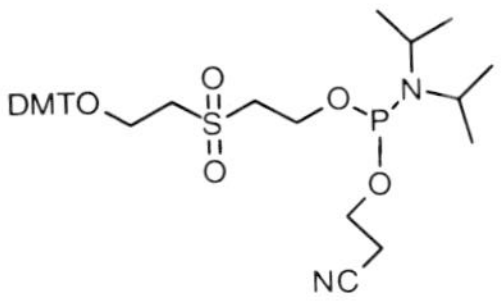

Figure 7. Monomer used to add a 5′-phosphate or a 3′-phosphate to a synthetic oligonucleotide.

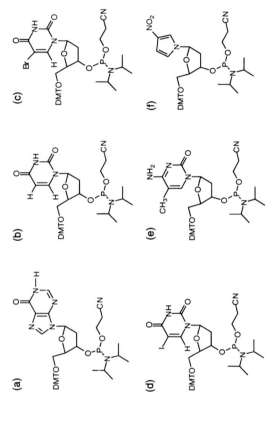

Figure 8. Modified nucleoside phosphoramidite monomers. (a) Deoxyinosine – universal base used at degenerate sites; (b) deoxyuridine – substrate for uracil DNA glycosylase; (c) 5-bromodeoxyuridine and (d) 5-iododeoxyuridine – for UV crosslinking of DNA to proteins; (e) 5-methyl deoxycytidine – uncommon base in DNA; (f) 3-nitropyrrole phosphoramidite – universal base used at degenerate sites.

2.6 Modified nucleosides

Modified nucleosides can be introduced into synthetic oligonucleotides during solid-phase synthesis, and several of these are illustrated in *Figure 8*. Deoxyinosine (*Figure 8a*) is widely used as a universal nucleoside (base) as it forms stable base pairs with all other Watson–Crick bases (with the

'PCR stoppers': (a) C_{16} Phosphoramidite monomer; (b) the structure of a PCR primer incorporating the C_{16} PCR stopper; (c) naphthosine phosphoramidite; (d) 1,4-anhydro-2-deoxy-D-ribitol in an oligonucleotide; (e) 1,3-propanediol in an oligonucleotide.

Chemically Labeled Primers

possible exception of G). It is widely available as a phosphoramidite. Recently a 3-nitropyrrole-2'-deoxynucleoside (*Figure 8f*) has been developed which has maximum base stacking and minimum hydrogen bonding properties [13]. It has been shown to be more suitable than deoxyinosine as a universal base. It is available from Glen Research as a phosphoramidite.

It is possible to synthesize oligonucleotides containing nucleosides bearing hydrophobic reporter groups such as biotin or fluorescein. However, care must be exercised when using such labeled oligonucleotides in PCR, particularly if they are added at or near the 3' end of a PCR primer. Undesirable interactions may occur between *Taq* DNA polymerase and the hydrophobic label, leading to inefficient PCR amplification. In addition, such monomers are extremely expensive.

2.7 'PCR stoppers' – synthesis of PCR products with single-stranded tails

'Nonnucleobasic' monomers incorporated into PCR primers

3 Suppliers of phosphoramidite monomers and other chemical labeling molecules

3.1 Amplimer custom synthesis

Amitof Biotech Inc, Appligene, Clontech, Cruachem, DNA International Inc., Genosys Biotechnologies Inc, Integrated DNA Technologies Inc., King's College School of Medicine and Dentistry, New England Biolabs, Operon, OSWEL (Oligonucleotide Service, Wellcome Trust), Pharmacia, Promega, R&D Systems Europe, Research Genetics, Severn Biotech Ltd.

3.2 Suppliers of phosphoramidites for amplimer synthesis

Advanced Biotechnologies, Applied Biosystems, Biometra, Boehringer Mannheim, Clontech, Cruachem, Genosys Biotechnologies Inc., GIBCO-BRL, Glen Research, Hoefer, Link Technologies, Pharmacia, Prime Synthesis, Sigma.

present an impassable barrier to primed synthesis by *Taq* DNA polymerase [14, 15]. When extended as PCR products these oligonucleotides no longer serve as templates for the polymerase beyond the insertion sites of the modified intermediates. They thereby produce single-stranded tails on amplification products. These tails can be of any chosen nucleotide sequence; there is therefore a vast number of possible alternatives which can act as specific partners of binding pairs. The other partner of any binding pair would be the complementary sequence to that of the tail. Solid-phase capture and subsequent hybridization detection of amplicons is therefore possible without the need to denature the PCR product. Examples of modified monomers that can be used to insert 'PCR stoppers' into PCR primers are shown in *Figure 9*. The naphthosine and C_{16} phosphoramidites can be obtained from Genosys Biotechnologies Inc.

Table 2 summarizes the information on common chemical labels.

Acknowledgment

T.B. is grateful to the Royal Society of Edinburgh for a Caledonian Research Fellowship.

Table 1. Approximate wavelengths of absorption and emission

Dye	Absorbance wavelength (nm)	Emission wavelength (nm)	Emission color guide
5- or 6-FAM[a]	500	525	Blue
TET	520	545	Green
JOE	525	555	Green
HEX	540	565	Green/yellow
TAMRA	555	580	Yellow
ROX	585	610	Red

[a]Fluorescein.

Table 2. Common chemical labels and their use in synthetic oligonucleotides

Label	Position	Conjugation method	Applications	Detection methods	Suppliers	Reference
Biotin	Single or multiple addition; internal or end labeling	Amidite	Capture or detect, *in situ* hybridization	(Strept)avidin or antibody; colorimetric or chemiluminescent	Most phosphoramidite suppliers	3, 16, 17
Digoxigenin	5'-end single addition	Active ester	Capture or detect, *in situ* hybridization	Antibody; colorimetric or chemiluminescent	Boehringer-Mannheim	4–6
Dinitrophenol	Single or multiple addition	Amidite	Capture or detect, *in situ* hybridization	Antibody; colorimetric or chemiluminescent	Cruachem, Clontech, some others	7
Fluorescein and derivatives FAM, HEX, TET, etc.	5'-end single addition	Amidite or active ester	Capture or detect, *in situ* hybridization, DNA sequencing	Antibody; fluorescence microscopy, laser fluorescence	Many phosphoramidite suppliers for fluorescein; Applied Biosystems for others	8, 10, 11
Rhodamine and derivatives – ROX TAMRA, etc.	5'-end single addition	Active ester	*In situ* hybridization, DNA sequencing	Fluorescence microscopy, laser fluorescence	Many phosphoramidite suppliers for rhodamine; Applied Biosystems or others	8, 11
Aminohexyl	5'-end single addition	Amidite	To couple to active esters (also for enzyme coupling)		Most phosphoramidite suppliers	18–20

Tetraethylene or hexaethylene glycol	Between oligonucleotide and label	Amidite	Spacer	Many phosphoramidite suppliers	21
Thiohexyl	5'-end single addition	Amidite	Enzyme coupling	Most phosphoramidite suppliers	1–4
5'-phosphate (3'-phosphate)	5'-end single addition (3'-end single addition)	Amidite	5'-phosphate for subsequent ligation (3'-phosphate not for PCR primers)	Most phosphoramidite suppliers	22
Inosine	Multiple internal addition	Amidite	Degenerate base for ambiguities in codons	Most phosphoramidite suppliers	23
Universal nucleoside	Multiple internal addition	Amidite	Degenerate base for ambiguities in codons	Glen Research	13
Bromo dU	Multiple internal addition	Amidite	UV catalyzed protein crosslinking	Most phosphoramidite suppliers	
PCR stoppers	Single internal addition	Amidite	To produce 5' single-stranded tail in PCR products	Genosys Biotechnologies Inc.	14, 15
Deoxyuridine	Multiple internal addition	Amidite	PCR product carry-over prevention (Chapter 9); directional amplicon cloning (Chapter 13)	Most phosphoramidite suppliers	25–27

Chemically Labeled Primers

Chapter 8 PROTOCOL OPTIMIZATION AND REACTION SPECIFICITY

– S.J. Powell

PCR is usually carried out to fulfill one of the following criteria:

1. Sample screening;
2. Target analysis;
3. Product preparation.

For routine screening purposes in which suitable conditions have already been established, optimization of the procedure is unnecessarily time consuming. However, for preparative PCR it is often necessary to enhance the production of a specific product. This may be in terms of yield, to reduce what may be a significant background of nonspecific products or minimize PCR-induced mutations.

This chapter aims to outline the methods of optimization that may be applied to the factors that influence the efficacy

impractical or unnecessary to optimize every PCR parameter. In fact, PCRs are usually carried out to fulfill only one of the four criteria (see *Table 1*). Identifying the main criterion allows us to focus our attention on specific parameters.

1 Reaction components

1.1 Template

The ideal template for a PCR is free from contaminants such as nucleases. Methods of template preparation are summarized in Chapter 4; Chapter 1, *Table 1* and Chapter 4, *Table 9* describe the relationship between initial template concentration and final product copy number. Efficacy of amplification may also be affected by the base composition of the template, but although secondary structure and sequence are

of a PCR, as outlined in *Table 1*. *Table 2* lists frequent laboratory applications of PCR, and indicates the relative importance of each of these factors. Some of the optimization procedures described here are expensive and time consuming; it is recommended that elementary parameters are changed before more elaborate steps are taken.

PCR optimization kits are available commercially from Advanced Biotechnologies, Biogene, Boehringer Mannheim, Invitrogen and Stratagene. These can provide a practical starting point; however, due to the intrinsic complexity of PCR, no one kit can cover all possible optimization strategies.

Individual elements of a PCR protocol may be optimized to generate a significant improvement in target production. Procedures based on the use of orthogonal arrays [1] and Taguchi methods [2] can be used to optimize several parameters simultaneously. These approaches rely on the accurate quantitation of product throughout optimization; thus they are time consuming and costly. It is generally

unique to each individual template, steps may be taken to relieve any problems encountered.

1.2 Buffer composition

Optimum buffer composition varies according to the enzyme and the PCR application. Tris–HCl, Tris–acetate, TAPS(Tris-[hydroxymethyl-methyl-amino-propanesulfonic acid) or Tricine may be used as the buffering agent. Most PCRs are carried out in the pH range 8.3–8.8, although for some applications more extreme pH values are effective, for example long amplicons are more readily amplified in the pH range 9–9.5. Mg^{2+} concentration can severely affect the efficiency of PCR as a consequence of its complexing with dNTPs and the Mg^{2+} requirement of the enzyme. An excess of Mg^{2+} results in increased nonspecific priming whereas too low Mg^{2+} levels reduce product yield. Optimum Mg^{2+} concentration should be attained empirically by titrating in 0.5 mM increments between 0.5 mM and 5 mM. The presence of monovalent cations (K^+ in most cases) can occasionally cause secondary structure formation in G-rich

Protocol Optimization

regions. This can lead to enzyme stalling and hence failure of the PCR – an effect which may be overcome by omitting KCl at the expense of enzyme stability [3].

Changing one, some or all of these buffer constituents may help to optimize amplification. *Table 3* summarizes buffer components and their effects.

Buffers may also be supplemented with auxiliary components known as PCR enhancers. This provides a convenient approach to limited optimization. However, the mode of action of these additives is not completely understood and the degree of success varies for each specific PCR. *Table 4* lists enhancers and their recommended working concentrations.

1.3 Choice and concentration of polymerase

Table 5 lists enzymes in relation to the PCR optimization criteria discussed. Generally, optimum enzyme concentrations lie in the range of 0.005–0.025 units μl^{-1}. Higher

Thermal cycling parameters have a significant effect on the four amplification properties outlined in *Table 1*. Generally, three-step cycling profiles establish a robust basis for optimization. This is described in *Table 6*.

2 Thermal cycling

When optimizing PCR it is important to ensure complete thermal equilibrium of the reaction mix. Accordingly, reaction volume (including oil or wax layer) and tube wall thickness are critical variables to consider when setting up cycling profiles. Most reactions are performed in 25–100 μl although 5–10 μl volumes are reported to be successful in glass capillary tubes where cycles of < 30 sec are possible due to rapid thermal equilibration [4]. Reactions are usually carried out in 0.5 ml or 0.2 ml reaction tubes. Several manufacturers supply thin-walled tubes (Chapter 2, *Table 1*) which can provide a distinct improvement in PCR efficiency over conventional tubes.

concentrations may cause an increase in nonspecific product generation. For further information, refer to suppliers' recommendations.

1.4 Primers

Primer design has an important influence on the specificity and efficiency of amplification, with respect to the following properties, which will be unique to each amplimer:

1. Base composition;
2. Length;
3. Chemical modifications;
4. Purity.

These factors will influence the generation of amplimer secondary structures and primer-dimer formations, and therefore should be taken into account when optimum reaction conditions are required. The concentration of primers added to a reaction will also influence yield, and it may be necessary to titrate the final concentrations in the range 0.1–1 µM.

2.1 Denaturation

It is critical that the first denaturation step of a PCR runs to completion, such that the strands of template DNA are entirely separated. In the case of complex genomic DNA or supercoiled plasmids, denaturation may have to be performed at 100°C for several minutes. Generally, 94–96°C for 2–3 min is sufficient. After initial denaturation, a slightly lower temperature of 92–95°C will be adequate for the remainder of the PCR.

2.2 Hot-start PCR

Assembling all the components of a PCR increases the potential of mispriming before thermal cycling has been initiated. In the hot-start technique, initial denaturation is performed in the absence of polymerase or primers. The temperature of the reaction mix is then maintained at 70–90°C until all components are combined. Thus enzyme and primers never coexist below the annealing temperature. Nonspecific priming is minimized and specificity and yield are increased.

Protocol Optimization

Hot-start PCR is simplified by using wax beads (Chapter 2, *Table 1*). The PCR reaction mix is assembled without enzyme or primers and aliquoted into reaction tubes. A wax bead is then added to each tube. Melting at 70–80°C, followed by cooling creates a solid plug on to which the remaining components are introduced. Thermal cycling is then initiated. The denaturation step melts the wax and causes the reaction components to mix. The presence of the wax does not impair the reaction.

2.3 Annealing

Calculating the temperature at which primers anneal specifically to a template is fundamental to efficient amplification (see Chapter 6 and ref. 5). However, the T_m of any given primer should be applied only as an approximate empirical indication of annealing temperature. As a rule of thumb, lowering the annealing temperature from the calculated T_m will increase the likelihood of nonspecific amplification. As the temperature is increased through the calculated T_m specificity will increase and yield will decrease.

cases where the target is present in extremely low copy number or absolute specificity is essential, nested PCR is often the most efficient method of optimization. Nested primers annealing internally to the first primers are used in a secondary PCR to amplify primary reaction products, thus producing a slightly smaller amplification fragment. The considerable increase in sensitivity and specificity achieved by use of nested PCR is due to the unlikely event of both the first and the second primer pairs having sufficient homology to nontarget sequences to amplify spurious fragments. This effect can also be attained, though to a lesser degree, when one nested primer is used in conjunction with a primary amplimer.

The major drawback of this approach for diagnostic applications is the increased risk of contamination due to the high sensitivity of nested PCR. This problem can be overcome by employing single-tube nested PCR which uses two primer pairs with different annealing temperatures, the primary pair having a higher T_m than the nested pair. Nested product is therefore only amplified in the latter stages of the cycling when the annealing temperature is reduced.

2.4 Polymerization

The extension rates of thermostable polymerases are between 2 and 4 kbp min^{-1}. Hence complete strand synthesis of 200 bp–5 kbp amplicons should be accomplished in less than 3 min (assuming other parameters have been optimized). For longer targets, extension steps may be anywhere up to 20 min long. Moreover, adding polymerization time extensions of 15–20 sec cycle^{-1} (using the incrementation facility available on some thermal cyclers) can improve specificity, yield and length of such products (see Section 4). In general, incrementation will enhance yield when enzyme becomes limited or during amplification plateaux.

3 Nested PCR

Although not necessarily the reason for optimization, specificity can have a direct influence on amplicon yield and length. Amplimers of 20–30 bp usually provide sufficient target specificity to fulfill a given PCR function. However, in for example, amplification of large genomic fragments. The following factors should be considered:

4 Long PCR

Routine PCR usually involves the generation of products less than 5 kb in length. This poses no great problem using conventional PCR protocols. Occasionally one may require the generation of much larger products (10–30 kb) for use in, for example, amplification of large genomic fragments. The following factors should be considered:

Template

The efficacy of long-range PCR is directly related to the complexity and purity of the starting material from which the target of interest is to be derived. For example, amplification from crude genomic DNA requires more stringent protocol optimization than amplification from a λ clone.

Denaturation

Total denaturation of target sequences is essential for successful amplification. This may be achieved by the

Protocol Optimization

addition of cosolvents and/or higher denaturation temperatures for a shorter period of time.

Specificity

Because of the extended nature of the target, shorter nonspecific fragments may interfere with long-range amplification. Less enzyme and lower salt concentrations will enhance specificity.

DNA damage

Long stretches of DNA are more susceptible to depurination and deamination at high temperatures. The introduction of cosolvents such as dimethylsulfoxide (DMSO) and/or glycerol [6] into the PCR mix decreases the temperature and step time necessary for complete denaturation, thus minimizing DNA damage. In some instances two or more PCR additives can give significant improvement. Depurination also occurs more readily at low pH [7]; a buffer with a low pH temperature coefficient such as Tricine reduces this risk.

Enzymes

Certain thermostable polymerases are more efficient than others when amplifying long amplicons (see *Table 5*). *Taq*, *Tth*, *Tli* (exo$^-$) and *Tbr* polymerases are all capable of producing PCR fragments of at least 12 kb, and using optimized buffer and cycling conditions can generate products in excess of 20 kb [6]. A combination of a polymerase containing no exonuclease activity with a low concentration of a polymerase with 3′–5′ exonuclease activity has been reported to enhance yield and product length further still [8]. It is proposed that the proofreading activity of the exo$^+$ enzyme is sufficient to remove mismatched bases incorporated by the exo$^-$ polymerase, thereby preventing stalling of the major component enzyme. Optimum ratios of exo$^-$ to exo$^+$ are in the range 10–600-fold but vary greatly with enzyme combination and PCR conditions. *Table 7* gives enzyme combinations cited for long PCR.

Table 1. Optimization for the amplification of fragments up to 5 kbp in length

Property of amplification to be optimized	Reason for optimization	Recommendations
Specificity	Eliminating nonspecific products	Increase annealing temperature Change Mg^{2+} concentration Change primer composition Titrate number of cycles Change polymerase type Add an appropriate PCR enhancer
Fidelity	Reducing error rate	Decrease number of cycles Increase template concentration Change polymerase type Remove any PCR enhancers
Yield	Increasing quantity of final product	Increase number of cycles Increase template concentration Change polymerase type Decrease annealing temperature Increase extension time for longer products
Length	Amplification of extensive target sequence	Increase extension time Change polymerase type Change Mg^{2+} concentration Decrease denaturation time Increase template concentration Change pH

For longer products, see Section 4.
The recommendations for optimization are listed in order of importance.

Table 2. Relative importance of optimization with respect to selected laboratory PCR applications

PCR applications	Specificity	Fidelity	Yield	Length
ARMS, ASA, ASP, PASA	++++	++	+	++
Alu-PCR	++++	+	+	+++
cDNA isolation	++++	++++	+++++	+++
Chromosome localization	++++	+	+	+
Construct analysis	++++	+	+	+
Cycle sequencing	++++	++	++	++
Degenerate PCR	+	+++	+++	+++
DNA fingerprinting	++++	++	++	++++
Gene isolation	++++	++++	++++	+++++
Gene walking	++++	+++	+++	+++++
Genetic screening	++++	++	+	++
Library screening	++++	+	+	+
Mutagenesis	++	++++	+	++
Subcloning	++	++++	+	++
RAPD	+	+	+++	++
Synthetic gene construction	+	++++	++++	++
YAC/cosmid mapping	++++	+	++	++

Table 3. Buffer components and properties

Buffer component	Comments	Optimization
Mg^{2+}	Essential for enzyme activity Influences template denaturation temperature Complexes dNTPs Excess Mg^{2+} causes an accumulation of nonspecific products Insufficient Mg^{2+} reduces product yield, sometimes to detection limits	Use a final concentration of between 0.5 mM and 5 mM Mg^{2+}. Titrate in 0.5 mM steps. Fine tune in 0.2 mM increments.
dNTPs	Optimal concentration is dependent on: product size primer concentration Mg^{2+} concentration reaction conditions nucleotide modifications	Decrease the final dNTP concentration from 100–200 mM to 20–40 mM for low cycle reactions
KCl	Monovalent salt increases activity of thermostable polymerases by 40–60% Excess KCl may inhibit polymerase activity Absence increases polymerase processivity	Dependent on polymerase (see Table 5)
$[NH_4]_2SO_4$	Enhances activity of specific polymerases	Dependent on polymerase (see Table 5)
Buffering agent	Tricine (300 mM) has been described as an effective alternative to Tris–HCl or Tris–acetate for longer products	See refs 6 and 9
pH	Excessively low pH levels cause nucleic acid degradation The fidelity of Taq polymerase increases as pH is decreased towards 5.0	Dependent on buffer composition

Table 4. PCR enhancers

Enhancer	Theoretical mode of action	Final concentration	Reference
DMSO	Destabilizes inter- and intrastrand reannealing	3–10%	6, 10–16
Glycerol	Improves denaturation Stabilizes polymerase in solution	10–15%	6, 17
PEG 6000	n.i.	5–15%	11
TMAC	Increases stringency of primer–template interactions	10–100 μM	18
Formamide	Increases denaturation	5–10%	14, 19–21
Tween 20/Nonidet P40	Reverses inhibitory effects of ionic detergents on polymerase	0.1–2.5%	14, 20
7-deaza dGTP	Weakens G–C interactions	75% of total dGTP	21
E. coli SSB protein	Enhances stability of ssDNA	5 μg ml^{-1}	22
Perfect Match (Stratagene)	n.i.	1 unit	23
Gene 32 protein (Pharmacia)	Enhances polymerase activity	1 nM	24

Abbreviations: DMSO, dimethyl sulfoxide; n.i., no information; PEG, polyethylene glycol; SSB, single-stranded binding protein; TMAC, tetramethylammonium chloride.

Table 5. Choice of polymerase to suit optimization criteria

DNA polymerase	Optimization properties	Requirements	Recommendations			
			Specificity	Fidelity	Yield	Length
Pfu (cloned)	3′–5′ exonuclease activity Low error rate	1.5–8 mM Mg 10 mM (NH$_4$)$_2$SO$_4$	++	++++	+++	+++
Pfu (exo⁻)	No exonuclease activity High temperature polymerization	1.5–8 mM Mg 10 mM (NH$_4$)$_2$SO$_4$	+++	+	+++	++
Psp	3′–5′ exonuclease activity	1.5–8 mM Mg 10 mM KCl 10 mM (NH$_4$)$_2$SO$_4$	++	+++	++	++
Psp (exo⁻)	No exonuclease activity High temperature polymerization	1.5–8 mM Mg 10 mM KCl 10 mM (NH$_4$)$_2$SO$_4$	++++	+	++	++
Taq	5′–3′ exonuclease activity	1–4 mM Mg 50 mM KCl	++	++	++	+++
Taq (N-terminal deleted)	No exonuclease activity High temperature polymerization	2–10 mM Mg 10 mM KCl	+++	+	+++++	+++
Tbr	5′–3′ exonuclease activity		++	++	+++	++

Continued

Table 5. Choice of polymerase to suit optimization criteria, *continued*

DNA polymerase	Optimization properties	Requirements	Specificity	Fidelity	Yield	Length
Tfl	No exonuclease activity	10–15 mM Mg 5–10 mM KCl	++	+	++	+++
Tli	3′–5′ exonuclease activity High temperature polymerization	2–8 mM Mg 10–50 mM KCl 10 mM (NH$_4$)$_2$SO$_4$	+++	+++	++	+++
Tli (exo⁻)	No exonuclease activity High temperature polymerization	2–8 mM Mg 10 mM KCl 10 mM (NH$_4$)$_2$SO$_4$	+++	++	+++	+++
Tma	3′–5′ exonuclease activity		++	+++	++	++
Tth	5′–3′ exonuclease activity		+++	++	++	++++

Table 6. Optimization properties of thermal cycling parameters

Cycling parameter	Optimization properties (length)	Optimization properties (yield)	Optimization properties (fidelity)	Optimization properties (specificity)
Denaturation temperature	94–96°C for products > 3 kbp		90–94°C to minimize DNA damage	
Denaturation time	5–30 sec for products > 3 kbp			
Annealing temperature		Decrease from calculated T_m to increase percentage of primer binding		Increase from calculated T_m
Annealing time		2–3 min to maximize primer binding	5–30 sec to reduce non-specific priming	
Polymerization temperature	Adjust to optimum recommended by polymerase manufacturer			
Polymerization time	Up to 20 min depending on enzyme stability Incrementation (available on some thermal cyclers – see Chapter 3)	Incrementation in plateau phase		
Cycle number		See Chapter 4, *Table 9*	Minimize by titration	

Protocol Optimization

Table 7. Examples of polymerase combinations for PCR of long amplicons

Major polymerase (exo⁻)	Minor polymerase (exo⁺)	Reference
Tth	*Tli*	6
Tth	*Psp*	6
Tth	*Tma*	6
Pfu (exo⁻)	*Pfu*	8
Pfu (exo⁻)	*Tli*	8
Pfu (exo⁻)	*Psp*	8
Taq	*Pfu*	6, 8
Taq	*Tli*	8
Taq	*Psp*	8

Chapter 9 **CONTAMINATION AVOIDANCE** – C.R. Newton

PCR sensitivity can be a potential problem or even a major limitation, since a false positive signal may arise from very low levels of contamination with amplicons from previous reactions. Although the physical separation of pre- and post-PCR manipulations (see Chapter 2) should reduce this risk dramatically, there are means of making carry-over contamination unamplifiable. These comprise:

1. Irradiation by UV light;
2. Gamma irradiation;
3. Restriction enzyme digestion;
4. DNase/exonuclease digestion;
5. Photoinactivation after isopsoralen treatment;
6. Photoinactivation after psoralen treatment;
7. Hydroxylamine hydrochloride treatment;
8. Substitution of dUTP for dTTP during PCR, and pre-PCR treatment with uracil DNA glycosylase (UDG);
9. Post-PCR alkali treatment after PCR using amplimers with 3′ ribose residues;
10. Sodium hypochlorite treatment of surfaces.

The choice of decontamination method should be influenced by whether the laboratory is already engaged in PCR or is to embark on PCR [1], since, for example, the introduction of dUTP substitution and UDG treatment will be of no benefit in an already contaminated environment; however, psoralen treatment will be efficacious. *Table 1* compares decontamination methods for PCR reaction mixtures.

RT-PCR presents a potential additional problem. The RNA template preparation may be contaminated not only by nucleic acids of external origin but also by genomic DNA from the same organism. In many instances this type of contamination can be identified and even tolerated, since amplimers may be designed that would span an intron–exon junction and give a larger amplicon than that derived from reverse-transcribed RNA [24]. However, this presupposes detailed knowledge of the structure of the gene and that the gene is not intronless, nor is it a processed pseudogene. *Table 2* compares decontamination methods for RT-PCR and *Table 3* compares decontamination methods for laboratory surfaces.

Table 1. Decontamination methods for PCR reaction mixtures

Method	Principle	Comments	Reference
UV irradiation	Amplimers and nucleotides less susceptible to dimerization between neighboring pyrimidine bases	Target DNA added after treatment Efficiency depends on length and sequence of contaminating DNA, therefore effective dose should be determined empirically	2, 3, 4
Gamma irradiation	Amplimers and nucleotides less susceptible to damage by free radicals formed from ionization of water	Target DNA added after treatment Efficiency depends on length and sequence of contaminating DNA, therefore effective dose should be determined empirically	5
Restriction enzymes(s)	Digests contaminating amplicons into fragments	Target DNA added after treatment Efficiency in PCR buffer depends on choice of restriction enzyme(s) – see also Chapter 14, *Table 1*	6, 7, 8

Nuclease	Degrades contaminating amplicons	Target DNA added after treatment (except exo III protocol [9])	6, 9, 10
Isopsoralen/UV irradiation	Contaminating amplicons modified by treatment can no longer function as template for PCR due to formation of photochemical adducts with DNA	Leaves modified amplicons single-stranded, therefore available for hybridization Isopsoralens can be included in the reaction mixture prior to PCR with UV irradiation post-PCR	11–14
Psoralen/UV irradiation	Treated contaminating DNA cannot be denatured	Modified amplicons are cross-linked double-stranded	1, 15, 16
Hydroxylamine	Modification of cytosine residues	Suitable only when amplicon detection is required after PCR (e.g. gel electrophoresis/EtBr staining)	17
dUTP/UDG	Substitution of dUTP for dTTP during PCR renders amplicon susceptible to UDG digestion prior to initiation of subsequent PCRs	Some restriction enzymes will not cleave after dUTP substitution [18,19] Some polymerases will not incorporate dUTP efficiently [20], see also Chapter 5 Not recommended where contamination already exists for current targets and primer pairs Annealing temperature of thermal cycle should not exceed 55°C to avoid residual UDG activity degrading newly synthesized amplicons [21]	1, 13, 14, 18, 21, 22

Continued

Table 1. Decontamination methods for PCR reaction mixtures, *continued*

Method	Principle	Comments	Reference
		For downstream cloning must use *ung*⁻ (UDG-deficient) host	22
dUTP-substituted amplimers	Primers incorporated into amplicons are UDG digested prior to initiation of subsequent PCRs	Annealing temperature of thermal cycle should exceed 55°C to avoid residual UDG activity degrading amplimers and newly synthesized amplicons [21] Primers added after UDG treatment	14, 23
3′ ribose amplimers	Primers incorporated into amplicons are NaOH digested prior to initiation of subsequent PCRs		

EtBr, ethidium bromide.

Table 2. Decontamination methods for RT-PCR

Method	Principle	Comments	Reference
Tagged reverse transcription primer	Only reverse-transcribed DNA is amplified using upstream primer and primer of same sequence as the tag	Only overcomes carry-over contamination if alternative tags are used for each experiment Efficient for prevention of genomic DNA contamination	25

Nuclease (RNase-free DNase)	Degrades contaminating DNA	Reverse transcriptase added after treatment	26, 27
Restriction enzyme(s)	Digests contaminating amplicons into fragments leaving reverse-transcribed ssDNA intact	Restriction digestion is final step prior to PCR, therefore contamination from all sources is removed Efficiency in PCR buffer depends on choice of restriction enzyme(s) – see also Chapter 14, *Table 1*	28
dUTP/UDG	Substitution of dUTP for dTTP during PCR renders amplicon susceptible to UDG digestion prior to initiation of subsequent PCRs	Some restriction enzymes will not cleave after dUTP substitution [18, 19] Some polymerases will not incorporate dUTP efficiently [20], see also Chapter 5 Not recommended where contamination already exists for current targets and primer pairs Annealing temperature of thermal cycle should exceed 55°C to avoid residual UDG activity degrading newly synthesized amplicons [21] Not efficient for prevention of genomic DNA contamination Single-tube method [30] eliminates potential contamination step of transfer from reverse transcription to PCR (analogous to hot-start) For downstream cloning must use *ung*⁻ (UDG-deficient) host	29, 30

ssDNA, single-stranded DNA.

Contamination Avoidance

Table 3. Decontamination methods for laboratory surfaces

Method	Principle	Comments	Reference
UV irradiation	Amplimers and nucleotides less susceptible to dimerization between neighboring pyrimidine bases	Efficiency depends on length and sequence of contaminating DNA, therefore effective dose should be determined empirically	31
Sodium hypochlorite	Oxidation of DNA causing extensive nicking	Effective even for small amplicons	32

Chapter 10 THE USE OF 5′ CHEMICALLY LABELED PRIMERS IN PCR

J. Grzybowski, F. McPhillips, D.J.S. Brown and T. Brown

1 Modified primers containing nonradioactive labels

The detection of PCR-amplified DNA can be achieved by the use of nonradioactively labeled primers (see Chapter 11). This precludes the hazardous, time-consuming and expensive introduction of radiolabels. In general, high detection sensitivity can be obtained by the covalent attachment of reporter groups to an oligonucleotide primer by means of one of the following systems:

1. *Fluorescent dyes* such as fluorescein and rhodamine and their derivatives, and Texas red. These are directly detectable by fluorescence microscopy or by laser-induced fluorescence on automated DNA sequencers.

2. *Haptens* such as fluorescein, biotin, digoxigenin and the 2,4-dinitrophenyl moiety (DNP). These are detectable colorimetrically or by chemiluminescence using antibody–enzyme conjugates and an appropriate enzyme substrate.

3. *Biotin*, which, in addition to haptens above, may be detected colorimetrically or by chemiluminescence using avidin (or streptavidin)–enzyme conjugates.

The attachment of most of the above labels to the 5′ ends of oligonucleotide primers can be carried out conveniently during automated solid-phase DNA synthesis. This methodology not only provides complete control over the entire synthesis procedure, resulting in the production of well-defined oligonucleotide products, but also allows the introduction of multiple reporter groups for increased detection sensitivity. However, the introduction of one or more of these generally hydrophobic moieties on to the 5' end of an oligonucleotide primer may have an adverse effect on the efficiency of subsequent PCR reactions. In many cases it is desirable to label both PCR primers with the same reporter group to increase detection sensitivity, or to label each primer with different chemical moieties for use in experiments involving both the capture and detection of PCR products.

In this chapter, we show that commonly used reporter groups attached to oligonucleotide primers can have a detrimental effect on the efficiency of PCR reactions. In the examples quoted *both* amplimers were chemically labeled at the 5' end to accentuate the inhibitory effect of the particular label. We also demonstrate the beneficial effects of incorporating hydrophilic spacers between the primer and the chemical label.

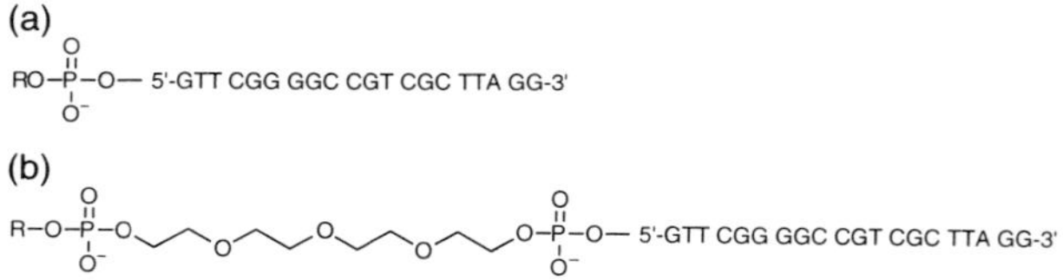

(a)
RO—P(=O)(O−)—O—5'-GTT CGG GGC CGT CGC TTA GG-3'

(b)
R—O—P(=O)(O−)—O—(CH₂CH₂O)₄—P(=O)(O−)—O—5'-GTT CGG GGC CGT CGC TTA GG-3'

Figure 1. The structure of PCR primer A, (a) attached directly to a reporter group (R) and (b) attached to a reporter group (R) via a hydrophilic tetraethyleneglycol spacer.

chemistry would allow, the effects of primer modification on PCR efficiency were tested by varying the numbers of labels attached to a single oligonucleotide primer. In the cases where increasing the number of labels resulted in a decrease in PCR efficiency, tetraethyleneglycol spacers were used to distance the labels from the oligonucleotide (*Figure 1b*). The results of the use of these modified amplimers are summarized in *Table 1*.

In conclusion, certain 5' modifications attached to oligonu-

2 Effect of modified oligonucleotide primers on PCR efficiency

Amplimers A (5′-GTTCGGGGCCGTCGCTTAGG-3′) and B (5′-CCCACGTGACCTGCCTCCA-3′) amplify a 389 bp fragment of a specific insertion sequence of *Mycobacterium avium* subspecies *paratuberculosis* [1, 2]. These were either labeled during synthesis by standard cyanoethyl phosphoramidite solid-phase chemistry, using commercially available biotin (Clontech), long-chain biotin (OSWEL/Link Technologies), fluorescein (Clontech) and 2,4-dinitrophenyl (Cruachem) phosphoramidites, or digoxigenin labeling was carried out by amino-derivatizing primers A and B at the 5′ end with the commercially available 'Aminolink 2' monomer (Applied Biosystems) prior to reaction with digoxigenin *N*-hydroxysuccinimide active ester (Boehringer Mannheim). Cholesterol and tetraethyleneglycol phosphoramidites were synthesized in our own laboratory and used as described previously [3, 4]. All primers were purified by reverse-phase HPLC. The structures of all labels used and their modes of attachment are illustrated in *Figures 1–6*. Where the

cleotide primers can have a detrimental effect on the efficiency of subsequent PCR experiments. This can sometimes be alleviated by the inclusion of one or more hydrophilic spacer molecules, or by reducing the number of labels attached to one primer. This will lead to a decrease

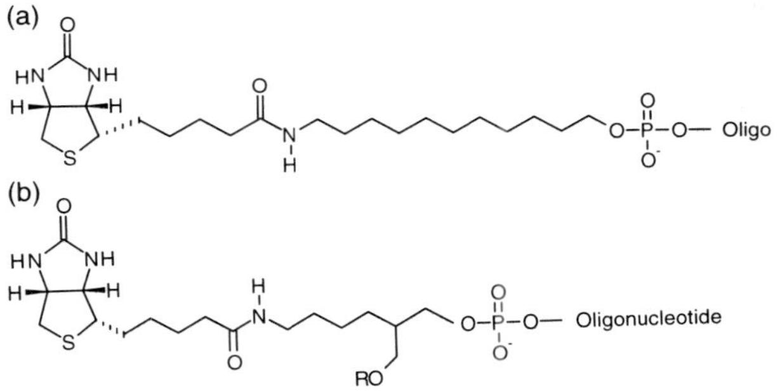

Figure 2. (a) Attachment of a long-chain biotin moiety to the 5′ end of a synthetic oligonucleotide. (b) Addition of biotin to the 5′ end of a synthetic oligonucleotide. This monomer is also used to add several biotin moieties to an oligonucleotide.

in subsequent detection sensitivity, but this may be balanced by allowing higher levels of amplification. The nature of 5'-label–enzyme interactions may differ according to the properties of the label, and either steric bulk or hydrophobicity may result in enzyme inhibition. This necessitates the individual optimization of each primer system. We are currently carrying out analogous PCR using labeled and unlabeled primers in combination. Preliminary data suggest that inhibition occurs, albeit to a lesser degree.

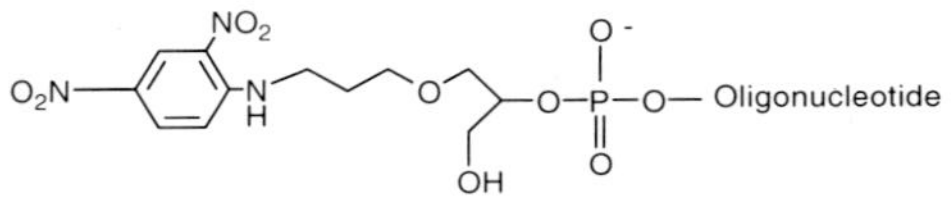

Figure 3. Structure of a synthetic oligonucleotide with a single DNP group added to the 5' end.

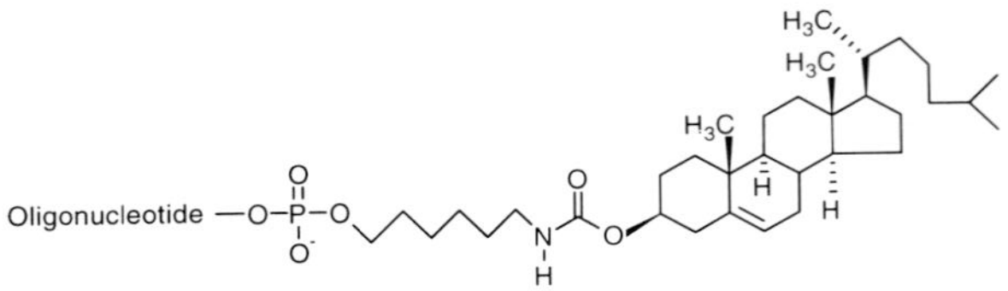

Figure 5. Attachment of a cholesterol moiety by a hexamethylene spacer to the 5' end of a synthetic oligonucleotide.

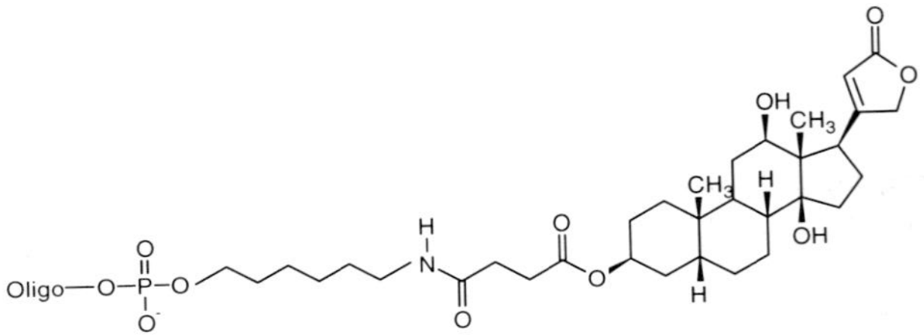

Figure 6. Structure of a synthetic oligonucleotide with digoxigenin attached at the 5' end by an aminohexyl spacer.

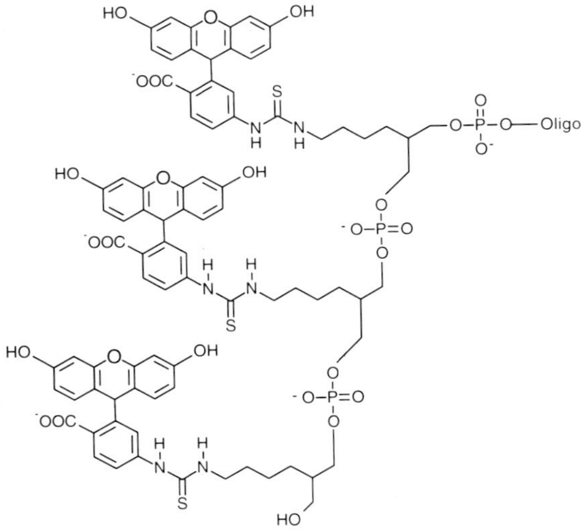

Figure 4. Multiple attachment of fluorescein to a synthetic oligonucleotide. The figure shows an oligonucleotide with three fluorescein molecules linked to the 5' end.

Acknowledgments

T.B. is grateful to the Royal Society of Edinburgh for a Caledonian Research Fellowship. The authors wish to thank Dr Karen Stevenson of the Moredun Animal Health Institute, Edinburgh for advice and the kind gift of IS900 cDNA. Patent applications relating to the methods described here are pending.

Table 1. Effect of modified primers on PCR efficiency

Label	Number of Tet-glycol spacers	Efficiency of PCR amplification[a]	Label	Number of Tet-glycol spacers	Efficiency of PCR amplification[a]
Unlabeled primers	0	+++	3 DNP	1	++
1 Biotin	0	+++	3 DNP TTTTT	1	+++
3 Biotin	0	+	8 DNP	0	–
3 Biotin	1	++	8 DNP	1	–
5 Biotin	3	+	8 DNP	2	–
1 Long-chain biotin	0	++	12 DNP	3	–
1 Cholesterol	0	–	1 Digoxigenin	0	–
1 Cholesterol	1	–	1 Digoxigenin	1	+
1 Cholesterol	2	–	1 Digoxigenin	2	+++
1 Cholesterol	3	–	1 Fluorescein	0	+
1 DNP	0	+++	3 Fluorescein	0	–
1 DNP	1	+++	3 Fluorescein	1	–
3 DNP	0	++	5 Fluorescein	2	–

[a] Key to PCR efficiency: +++, equivalent to unlabeled primers; ++, >50% but <90% efficiency of unlabeled primers; +, >0% but <50% efficiency of unlabeled primers; –, no amplification.
Tet-glycol, tetraethyleneglycol.

Chapter 11 **DETECTION OF PCR AMPLIFIED PRODUCTS** –
L.J. MacCallum

Unless conditions have been optimized to maximize amplicon yield (Chapter 8), specific product(s) from a PCR may not be easily detectable. Moreover, not all PCRs yield a single product; nonspecific priming may generate superfluous products from which the target of interest is difficult to distinguish. Thus optimizing detection will enhance visualization of a specific amplicon. Before selecting a method for detecting PCR products, the following factors should be considered:

1. Purpose of the PCR;
2. Intended use of amplified products;
3. Versatility of the protocol – how easily can conditions be altered to optimize detection?
4. Sensitivity, specificity and reproducibility of the detection method;
5. Expense – can the technique be applied economically to routine assays? How labor intensive is the procedure?

This chapter focuses on the most user-friendly methods of amplicon detection and visualization. Basic techniques are summarized in *Table 1*; some of the commercial detection kits and apparatus that have evolved from such approaches are listed in *Table 2*.

1 Homogeneous detection of PCR products

Homogeneous PCR product detection (also known as single-tube or single-step detection) establishes the presence or absence of amplified sequences, but not the specificity of a PCR. Each reaction is set up with an internal detection system such as ethidium bromide (EtBr), an acridine dye that intercalates between the stacked bases of dsDNA helices. EtBr–dsDNA complexes fluoresce when illuminated with UV light. Since the ingredients of a PCR are essentially single stranded and the products of amplification are double stranded, adding EtBr as a component of a PCR results in an exponential increase in fluorescence with each cycle of amplification. Once PCR is initiated, the reaction tubes need not be reopened; this significantly reduces the risk of carry-over contamination. The presence of EtBr does not affect the specificity of the reaction, but because the assay is based on the production of dsDNA, primer–dimer formation will influence the intensity of fluorescence.

Measuring the incorporation of EtBr into PCR products
1. Before thermal cycling, add EtBr to the reaction mix at a final concentration of 4 μg ml^{-1}.
2. Observe EtBr fluorescence using a UV transilluminator. Suitable illumination wavelengths are listed in *Table 3*.

1.2 Homogeneous real-time analysis
By continually monitoring the production of dsDNA over the course of amplification, homogeneous detection systems may be used for both quantitative and kinetic PCR analysis. This is also a means of observing the effects of altered parameters; collation and analysis of real-time assay data will help to optimize conditions for future reactions.

1. Real-time analysis is best performed using camera or video equipment. Save an image of each reaction during each annealing/extension phase and continue monitoring over the complete course of the PCR [2].
2. The number of cycles necessary to produce detectable fluorescence is directly related to the number of target DNA copies in the original reaction mix and to the efficiency of the reaction (see Chapter 4, *Table 9*).

2 Gel detection of PCR products

Size separation through a gel matrix of agarose or polyacrylamide is traditionally the most popular means of

3. A fluorimeter is necessary for quantitative analysis.
4. Complex monitoring systems have been established for more accurate measurements. A spectrofluorimeter linked to an optical fiber will illuminate single PCRs and record fluorescence measurements [1].
5. For monitoring numerous amplifications simultaneously, a closed-circuit television camera will capture fluorescence images of PCRs in the sample block of a thermal cycler [2]. Use image-analysis software to digitize and manipulate results.

1.1 Homogeneous endpoint analysis

Homogeneous endpoint analysis will ascertain PCR efficacy. It is not applicable to quantitative PCR analysis.

1. Measure fluorescence before and after amplification. The accumulation of dsDNA will be evident by an increase in fluorescence.
2. Since unbound EtBr is weakly fluorescent, background is unavoidable. Always carry out a negative control.

detecting individual amplicons from a PCR. Samples are loaded into wells molded during gel preparation, and an electric field is used to separate the reaction products through the matrix. Visualizing the separation enables evaluation of product sizes and relative yields. Note that a band of the expected size is not absolute identification or characterization of a PCR product. Excision and elution of a fragment from the gel (Chapter 12), followed by reamplification of the eluate if necessary, should provide sufficient material for identification by diagnostic restriction endonuclease digestion, hybridization to a specific internal probe, or direct sequencing.

Sample preparation and loading

Remove an appropriate volume of each post-PCR and add loading buffer to 1× concentration. Loading buffers increase sample density to ensure accurate loading; they also contain dyes to enable sample migration to be observed. Recipes are given in *Table 4*. If an oil overlay has been used in the PCR, ensure that this does not contaminate samples by extracting with chloroform. Assemble the electrophoresis apparatus

Detection of Amplified Products

and fill the reservoirs of the gel tank with enough electrophoresis buffer to flood the sample wells. Recipes of popular electrophoresis buffers are given in *Table 5*. Clean each well thoroughly by flushing with buffer; lining each with a small quantity of loading buffer will identify any defects. Pipette the samples slowly into each well without overloading and start electrophoresis immediately.

Running the gel
Gel temperature should be uniform throughout electrophoresis as overheating may cause band distortion. Power requirements depend on agarose concentration, gel size and running buffer. Measure the distance between the electrodes of the electrophoresis tank and apply a voltage of $1-8 \text{ V cm}^{-1}$. For the most effective means of controlling the temperature of specific apparatus, consult the manufacturer's instructions. During electrophoresis, bubbles will rise from the electrodes and the dyes in the loading buffer will move towards the anode.

Product visualization by ethidium bromide staining
The most common means of visualizing gel-separated DNA

Product detection by amplicon labeling
The most sensitive means of amplicon detection is by the incorporation of label during polymerization. After gel electrophoresis, the products of the reaction can be visualized either directly from the gel or after transfer to a membrane support. This is discussed further in Section 4.

Product detection using internal labeled probes
The endogenous exonuclease activity of *Taq* DNA polymerase is exploited in a system where the generation of specific amplicons is detected by the simultaneous degradation of labeled sequence-specific probes [7–9]. These probes are added to the reaction mix before amplification, and do not inhibit the efficiency of the reaction. The result is a sensitive assay with adjustable specificity. Another method, using an internal probe to generate full-length labeled products which are detectable by gel electrophoresis is described in ref. 10.

Size analysis
Amplicon sizes can be ascertained by the relative electrophor-

bands is by staining with EtBr [3]. The sensitivity of this technique is approximately 0.1 ng mm^{-2} to a minimum of 5 ng when fresh stain is used. Stain fragments according to one of the methods described in *Table 6* and examine the gel on a UV transilluminator (*Table 3*). The target amplicon and any additional nonspecific products will normally be revealed as bright bands. If results are to be recorded, photograph the transilluminated gel. A gel documentation system (e.g. The Imager, Appligene) will enhance and document results with high resolution.

Product visualization by silver staining

The sensitivity of staining polyacrylamide-separated DNA fragments with silver nitrate in formaldehyde [4] is approximately 1 pg mm^{-2} when fresh, high-quality reagents are used [5, 6]. This technique provides a permanent record of detection, unlike EtBr fluorescence which is sequestered by polyacrylamide over time. The staining procedure may be optimized for different gel types and thicknesses, although instructions for commercial kits (e.g. Silver Stain Plus, BioRad) should be adhered to.

etic migration of at least one set of DNA molecular weight markers. Electrophorese the markers of choice (which should reflect expected product size) in 1×loading buffer alongside samples in parallel gel lanes. PCR product sizes can be estimated by eye after staining or, for a more accurate calculation, by interpolation from a log plot of migration of marker fragments. Placing a ruler on or alongside the gel during photography will help with measurements and calculations. Size markers are available from most molecular biology product suppliers [11]. They are also suitable for approximate quantitation of products when a known marker mass is used.

2.1 Agarose gel electrophoresis (AGE)

Electrophoresis through an agarose matrix is the simplest and most cost-efficient means of estimating the efficiency of amplification and the size of PCR-generated fragments. Migration of DNA through agarose is ultimately dependent on:

1. Molecular weight of the fragment;
2. Concentration of agarose;

Detection of Amplified Products

3. Buffer composition;
4. Presence of EtBr.

Agarose gels are usually prepared as horizontal slabs using specialized casting apparatus. The preparation of agarose gels is summarized in *Table 7* [12]. *Table 8* shows the resolution ranges of percentage agarose matrices.

2.2 Polyacrylamide gel electrophoresis (PAGE)

Although PAGE is more expensive and time consuming than AGE, it is the recommended method of product detection when high resolution of DNA fragments over a wide range of sizes is required. PAGE gel and buffer specifications may be varied over a wide range to meet the specificity of separation required. The electrophoretic mobility of DNA through polyacrylamide is dominated by size; effective separation ranges are listed in *Table 9*. Note that the influence of charge on electrophoretic mobility becomes increasingly important as the pH decreases from 8.3. Polyacrylamide gels, which are usually run in a vertical compared to gel electrophoresis and EtBr or silver staining. Procedures and applications are summarized in *Table 13*.

Prior to blotting, gel electrophoresis and staining of an aliquot of the amplification reaction will establish product sizes, PCR efficiency, and may help to highlight and interpret any false hybridization results. This is particularly relevant before a dot blot or reverse dot blot since in these cases the products of the PCR are not separated.

Solid supports

Two classes of solid support are suitable for blotting:

1. *Nylon.* Nylon binds DNA irreversibly and is durable enough to withstand sequential hybridizations. The surface of nylon membrane may be modified with a positive charge to increase binding capacity.
2. *Nitrocellulose.* Nitrocellulose binds DNA through reversible hydrophobic interactions, therefore a slow leaching may occur throughout blotting. Nitrocellulose does not bind small fragments efficiently, and filters may become brittle and dry with overuse. However, levels of back-

plane between glass plates, may be denaturing or nondenaturing. Applications are listed in *Table 10*. Preparation is described in *Table 7* [12].

2.3 Denaturing gradient gel electrophoresis (DGGE)

DGGE is used to search for and identify mutations and polymorphisms [13]. The separation is based on the phenomenon that dsDNA molecules differing by a single base have slightly different melting (denaturation) properties [14, 15]. Samples are amplified from genomic DNA, then separated through polyacrylamide plus a gradient of denaturant. After electrophoresis, the gel is stained with EtBr or silver. Perpendicular and parallel DGGE are summarized in *Table 12*. A detailed description of DGGE procedures is given in refs 16 and 17.

3 Membrane detection of PCR products

Immobilization of DNA on to a solid support followed by hybridization to at least one internal probe enhances the sensitivity and characterization of specific product detection

ground hybridization are considerably less than those for nylon.

Most nylon and nitrocellulose membranes are manufactured with a 0.45 µm filter, although some have a decreased pore size. 0.20 µm membranes will retain fragments less than 300 nt. A selection of specialized membranes is listed in *Table 14*.

Probes

Probes for specific hybridization may be generated from either synthetic oligonucleotides or from existing sequences. Commercial kits for probe generation are widely available. Radionucleotides suitable for end-labeling and incorporation are listed in Chapter 14, *Table 4*. Of these, ^{32}P is the strongest emitter and therefore the most sensitive (0.02 ng mm^{-2} to a minimum of 10 fg DNA), but fresh label should be used routinely due to the short half-life of the isotope (see Chapter 14, *Table 5*). The addition of radiolabel to the 5′ and 3′ ends of probes is catalyzed by the enzymes polynucleotide kinase and terminal deoxynucleotidyl transferase, respectively [12]. Random primers may be used to incorporate

Detection of Amplified Products

radionucleotides into probes synthesized from a ssDNA or dsDNA template [18,19]. After hybridization, detection and visualization are by autoradiography, phosphorimaging or β-scanning.

Nonisotopic detection and visualization systems are especially valuable to laboratories processing large numbers of samples. These systems do not have associated biohazards or short half-life problems. Oligonucleotides labeled with enzymes (e.g. alkaline phosphatase, horseradish peroxidase), haptens (e.g. biotin, digoxigenin), fluorescent compounds (e.g. fluorescein, rhodamine) and chemiluminescent compounds (e.g. acridinium esters, luminol) all provide simple and safe methods of analysis. Nonisotopic labeling and detection systems are discussed in full in ref. 20.

3.1 Southern blot

In a Southern blot [21] of PCR products (see *Table 13*), multiple products may be detected by sequential hybridization (washing the solid support free of the existing probe,

ASOs allows simultaneous analysis of more than one allele [26, 27, 30–32]. For each membrane, the T_ms of the ASOs need to be similar to retain specificity for each hybridization. Reverse dot blot controls and procedures are summarized in *Table 15*.

4 Detecting labeled PCR products

Labeled PCR products are necessary for reverse dot blotting (Section 3.3), cycle sequencing [33], DNA protein binding/protection (footprinting) assays [34], examining microsatellites [35], DDRT-PCR [36, 37], *in situ* PCR [38, 39], and polymorphism/mutation detection by chemical cleavage [40]. They are detectable by endpoint reaction analysis or by gel electrophoresis.

Amplicons may be labeled by the direct incorporation of radioactive dNTPs, hapten-dNTPs (e.g. biotin-dUTP), or fluorescent dNTPs (e.g. fluorescein-dUTP) during a PCR. This is achieved by substituting a quantity of unlabeled dNTP for the corresponding labeled dNTP when setting up

then rehybridizing with a different probe), or by simultaneous hybridization (using a mixture of probes). Southern blotting controls and procedures are summarized in *Tables 15* and *16* [12].

3.2 Dot blot
A dot blot or slot blot is appropriate for simultaneous analysis of multiple PCR samples [22] (see *Table 13*). The technique is particularly suitable for high throughput assays; pairs of allele-specific oligonucleotides (ASOs) are routinely used in dot blot hybridizations to detect and analyze known allelic variations [23–27]. When ASO probes are used, hybridization conditions are determined by the T_m (Chapter 6). Controls and procedures for dot blotting are summarized in *Table 15*.

3.3 Reverse dot blot
Reverse dot blots are an alternative format to dot blots (see *Table 13*). They are advantageous when a large number of probes is required (e.g. for multiple mutations of a single gene) [28, 29]. Multiplex PCR incorporating several pairs of

the reaction. An alternative labeling technique is to utilize modified primers so that each amplicon inherits the functional moiety from its specific amplimers. Oligonucleotides may be 5′-end labeled by the enzyme polynucleotide kinase [12] or modified during chemical synthesis (see Chapters 7 and 10). Methods in which primers incorporate binding sites for triple helices and dsDNA-binding proteins into amplicons are described in references 41 and 42, respectively. Alternatively, the application of nonnucleosidic phosphoramidites during primer synthesis creates a barrier to further extension by *Taq* DNA polymerase (see Chapter 7). The outcome of the PCR is a nonamplifiable, single-stranded tail attached to each end of the amplicon. These tails can be specifically designed to label PCR products with unique ssDNA extensions; thus enabling capture by specific oligonucleotide probes.

4.1 Endpoint analysis using radiolabeled amplicons
Although semi-quantitative, measuring the incorporation of radiolabel into the products of a PCR is best confined to determining the presence or absence of product.

Detection of Amplified Products

1. After amplification, remove unincorporated dNTPs or labeled primers by spin dialysis or size exclusion (Chapter 12).
2. Measure residual radiolabel using a scintillation counter.
3. Count before and after thermal cycling. Compare values with negative control reactions.

4.2 Gel/membrane detection of labeled amplicons

Labeled PCR products may be detected and analyzed efficiently using gel electrophoresis. After separation (see Section 2), dry the gel under vacuum at 60–80°C (radiolabels only) or transfer fragments to a solid support (see *Table 18*).

Visualize bands by an appropriate detection method. Labeled DNA size markers are available from many suppliers of molecular biology products. Note that because all products become labeled during amplification, interpreting banding patterns may prove difficult if nonspecific fragments are of a similar size to target products. Fluorescent labels and automated detection methods circumvent this problem and enable accurate size analysis and quantitation of gel-separated fragments [43]. Systems have been developed for the simultaneous analysis of products from a multiplex PCR in which the primer pairs have been labeled with dyes that fluoresce at different wavelengths (e.g. 373 DNA Sequencer, Applied Biosystems).

Table 1. Techniques for detecting PCR products

Technique	Visualization and applications
Incorporation of EtBr into amplicons during synthesis (see Section 1)	UV transillumination homogeneous endpoint analysis homogeneous quantitative analysis homogeneous kinetic real-time analysis Gel electrophoresis, UV transillumination effectiveness of amplification estimation of product sizes
Incorporation of radiolabel into amplicons during synthesis (see Section 4)	Scintillation detection endpoint analysis Gel electrophoresis, transfer to membrane (optional), autoradiography, phosphorimaging or β-scanning effectiveness of amplification estimation of product sizes
Incorporation of nonradioactive label into amplicons during synthesis (see Section 4)	Gel electrophoresis, transfer to membrane, tag capture and detection effectiveness of amplification estimation of product sizes
Incorporation of fluorescent label into amplicons during synthesis (see Section 4)	Laser-induced fluorescence estimation of product sizes analysis of multiplex PCR genetic typing

Continued

Table 1. Techniques for detecting PCR products, *continued*

Technique	Visualization and applications
Agarose gel electrophoresis (see Section 2)	EtBr staining, UV transillumination effectiveness of amplification estimation of product sizes purification of amplicons Southern blot (see below)
Polyacrylamide gel electrophoresis (see Section 2)	EtBr staining, UV transillumination effectiveness of amplification estimation of product sizes purification of amplicons Silver staining effectiveness of amplification estimation of product sizes Southern blot (see below)
Denaturing gradient gel electrophoresis (see Section 2)	EtBr staining, UV transillumination detection and analysis of polymorphisms Silver staining detection and analysis of polymorphisms
Restriction endonuclease digestion (Chapter 14)	Gel electrophoresis, EtBr staining, UV transillumination product characterization purification of ready-to-clone amplicons RFLP analysis

Detection of Amplified Products

	PAGE, silver staining
	product characterization
	RFLP analysis
	HPLC (see below)
Southern blot (see Section 3)	Hybridization to specific probes
	PCR specificity
	estimation of specific product sizes
	product characterization
	confirmation of product identity
Dot blot (see Section 3)	Hybridization to specific probes
	PCR specificity
	allele typing
Reverse dot blot (see Section 3)	Hybridization of labeled PCR products to specific oligonucleotides
	PCR specificity
	allele typing (multiplex PCR)
High pressure liquid chromatography (HPLC)	UV detection
	effectiveness of amplification
	estimation of product sizes
	product purification
Sequencing	Autoradiography, phosphorimaging or laser-induced fluorescence (automated DNA sequencer)
	confirmation of product identity
	sequence analysis

Table 2. Commercially available amplicon/DNA detection kits

Kit	Technique	Features
VisiGel Separation Matrix (Stratagene)	Concentrated solution for generation of gel electrophoresis matrices Visualization by EtBr staining	Suitable for separating products < 1200 bp Tear-resistant gels High fragment resolution Allows DNA recovery from gel fragments
PhastSystem (Pharmacia)	Automated PAGE, staining and development	Precast gels High speed separation, staining and transfer for rapid screening PhastTransfer blotting unit available
DNA DipStick (Invitrogen), FastCheck (GIBCO-BRL)	Determination of post-PCR DNA concentration by colorimetric titration	Picogram sensitivity No radioactivity or EtBr involved
PCR ELISA (Boehringer Mannheim)	Digoxigenin-dUTP is incorporated into amplicons during synthesis Products are captured and immobilized by biotinylated probes Hybrids are detected with an anti-digoxigenin-peroxidase conjugate	Semi-quantitative 96-well microtiter plate format for high throughput
GEN-ETI-K DEIA Enzyme Immunoassay (Sorin Biomedica)	Products are captured and immobilized by biotinylated probes Hybrids are detected with a specific antibody followed by an enzyme-linked anti-species antibody	Nonradioactive Specific detection of a single target amplicon

SHARP Signal System (Digene Diagnostics Inc.)	Primers incorporate biotin into amplicons Products are hybridized to an RNA probe and immobilized on to a streptavidin-coated plate Hybrids are detected with an AP-conjugated antibody specific for DNA/RNA hybrids	Nonradioactive High sensitivity RNA/DNA hybrids are more stable than DNA/DNA hybrids Background hybridization is reduced by RNase treatment
CAPTAGENE-GCN4 (Pharmacia)	Primers incorporate biotin and the GCN4 recognition sequence into amplicons Products are captured by GCN4 and detected using a streptavidin–HRP conjugate	Colorimetric Nonradioactive High sensitivity 96-well microtiter plate format for high throughput
AMPLICOR PCR (Roche Diagnostic Systems)	Primers incorporate biotin into amplicons Products are captured and immobilized by a specific probe then detected with an HRP–streptavidin conjugate	Nonradioactive High sensitivity Supplied as a complete package for specific detection of a single target amplicon
*Quant*Amp (Amersham)	Amplicons labeled with a low-energy isotope are specifically captured by scintillant beads (fluomicrospheres) which fluoresce upon excitation by atomic decay Light emission measured by scintillation counting	Specific detection of target amplicons Quantitative May be performed in a 96-well microtiter plate format for high throughput
ECL Probe-Amp (Amersham)	Fluorescein-dUTP is incorporated into amplicons during synthesis Products are detected with a monoclonal anti-fluorescein–HRP conjugate	Nonradioactive High sensitivity (<0.5 pg)

Continued

Detection of Amplified Products

Table 2. Commercially available DNA detection kits, *continued*

Kit	Technique	Features
Electrochemoluminescence QPCR System 5000 (Applied Biosystems)	A voltage is applied across specifically captured, TBR-labeled PCR products Products are detected and quantified by analysis of light emission	Nonradioactive Quantitative Specific detection of target amplicons Labeling/detection assay formats can be varied depending on expected product yield
Cycle Sequencing Kit (Pharmacia), Cyclist *Taq* DNA Sequencing Kit (Stratagene), Cyclist Exo⁻ *Pfu* DNA Sequencing Kit (Stratagene), CircumVent Thermal Cycle DNA Sequencing Kit (New England Biolabs), *fmol* DNA Sequencing System (Promega)	Direct-cycle sequencing of amplicons using end-labeled primers or by incorporation of labeled nucleotides	Nanograms of template required No preparation of template prior to sequencing
CircumVent Phototope Kit and Phototope Detection (New England Biolabs)	Direct-cycle sequencing of amplicons using Vent$_R$ (exo⁻) DNA polymerase and a biotinylated primer	Nonradioactive Chemiluminescent detection Nanograms of template required
SILVER SEQUENCE DNA Sequencing System (Promega)	Direct-cycle sequencing of amplicons using *Taq* DNA polymerase and SILVER SEQUENCE labeled nucleotides	Nonradioactive Band visualization by silver staining Nanograms of template required

Abbreviations: AP, alkaline phosphatase; HRP, horseradish peroxidase; TBR, tris (2,2'-bipyridine) ruthenium (II) chelate.

Table 3. Fluorescence of EtBr in UV light (note that prolonged exposure to any wavelength of UV light will damage nucleic acids)

Illumination wavelength	Comments
254 nm	High sensitivity
	May cause photo-nicking of DNA or bleaching of visible fragments
312 nm	Maximum fluorescence
366 nm	Low sensitivity
	Low damage; good for preparative work

Safety note: UV irradiation is a powerful mutagen. It rapidly burns skin and has the potential to cause blindness by damaging the cornea. Always use complete body and face protection when working with UV light.

Table 4. Loading buffers (stock solutions) for gel electrophoresis

6 × loading buffer	Storage	Applications
0.25% Bromophenol blue 0.25% Xylene cyanol FF 40% (w/v) Sucrose in water	4°C	AGE and PAGE
0.25% Bromophenol blue 0.25% Xylene cyanol FF 15% Ficoll 400 (Pharmacia) in water	Room temperature	AGE
0.25% Bromophenol blue 0.25% Xylene cyanol FF 30% Glycerol in water	Room temperature	AGE and PAGE
98% Deionized formamide 0.025% Bromophenol blue 0.025% Xylene cyanol FF	4°C	Denaturing PAGE
0.25% Bromophenol blue 40% (w/v) Sucrose in water	4°C	AGE

These may be made up in water or 1 × electrophoresis buffer (*Table 5*). Abbreviations: AGE, agarose gel electrophoresis; PAGE, polyacrylamide gel electrophoresis.

Detection of Amplified Products

Table 5. Buffers for casting and running gel electrophoresis systems

Buffer	Working concentration	Stock solution (per liter)	Comments
Tris-acetate (TA)	1 × TA: 40 mM Tris-acetate	50 × TA: 48.4 g Tris base 11.4 ml glacial acetic acid	Store at 4°C Good for preparative gels The absence of EDTA reduces buffer interference with polymerase activity if agarose pieces are to be used in further amplifications
Tris-acetate-EDTA (TAE)	1 × TAE: 40 mM Tris-acetate 1 mM EDTA	10 × TAE: 48.4 g Tris base 11.4 ml glacial acetic acid 20 ml 0.5 M EDTA (pH 8.0)	Store at room temperature The cheapest and most commonly used buffer, but with the lowest buffering capacity For long electrophoresis runs, replacement or recirculation may be necessary to prevent buffer exhaustion
Tris-borate-EDTA (TBE)	0.5 × TBE: 45 mM Tris-borate 1 mM EDTA 1 × TBE: 89 mM Tris-borate 2 mM EDTA	10 × TBE: 108 g Tris base 55 g boric acid 40 ml 0.5 M EDTA (pH 8.0)	Store at room temperature Discard stock solution if a precipitate forms
Tris-phosphate-EDTA (TPE)	1 × TPE: 89 mM Tris-phosphate 2 mM EDTA	10 × TPE: 108 g Tris base 15.5 ml 85% phosphoric acid 40 ml 0.5 M EDTA (pH 8.0)	Store at room temperature High buffering capacity Unsuitable for product recovery by ethanol precipitation

Prepare all solutions using distilled water. Liquid concentrates of premixed buffers are also available commercially (e.g. BioRad).

Table 6. Methods of staining gel-separated PCR products with EtBr

Method of staining	Method of destaining	Comments
Before amplification, incorporate EtBr into the reaction mix at a final concentration of 4 µg ml^{-1}	N/A	Fragment mobility may be reduced by up to 15%
Incorporate EtBr into gel and electrophoresis buffer at 0.5 µg ml^{-1}	N/A	Fragment mobility may be reduced by up to 15%
0.5 µg ml^{-1} EtBr in electrophoresis buffer or distilled water, 30–60 min	Running buffer, 30 min	Destaining not always necessary
1–5 µg ml^{-1} EtBr in 0.5 M ammonium acetate, 30–60 min	Distilled water or 0.5 M ammonium acetate, 90–120 min	Destain in the dark

Prepare a 10 mg ml^{-1} EtBr stock and store in the dark at room temperature. Stock solutions are available commercially (e.g. *ultra*PURE 10 mg ml^{-1} ethidium bromide, GIBCO-BRL).

Abbreviations: EtBr, ethidium bromide; N/A, not applicable.

Table 7. AGE and PAGE preparation

	Agarose gel electrophoresis	Polyacrylamide gel electrophoresis
Raw materials	Agarose must be of DNA grade to be free from nucleases and proteases Specialist high-resolution agarose is available (e.g. NuSieve GTG, FMC Bioproducts) Low melting point (LMP) agarose is excellent for the efficient recovery of PCR products from agarose (see Chapter 12). LMP gels melt at a temperature of approximately 65°C, therefore it is recommended that electrophoresis is carried out in a cold room. All other running conditions are identical to those for standard gels	Prepare acrylamide as a 29% acrylamide/1% N,N'-methylenebisacrylamide stock solution. Dissolve in distilled water at 37°C and store at 4°C. Do not expose to light. Stock solutions are available commercially (e.g. EAS*igel*, Scotlab).
Buffers	See *Table 5* Use fresh buffer for each electrophoresis run If possible, recirculate the buffer between the reservoirs of the electrophoresis tank during long separations	1 × TBE (see *Table 5*) Always use buffer from the same stock used in gel preparation to prevent migration artifacts
Gel casting	Make sure all equipment is clean A cold lab will speed up the casting process and increase the transparency of the gel Combs may be inserted before or after pouring (but before setting) Mix the agarose to the desired concentration (%w/v) with 1 × electrophoresis buffer in a container of at least three times the volume of the final agarose solution. Boil until all the agarose has dissolved. Beware if heating in a microwave oven; superheated molten agarose can boil over spontaneously	Make sure all equipment is clean Treating the inner surface of one plate with silicone (e.g. Sigmacote, Sigma) will prevent the gel from tearing when the plates are separated The choice of spacers will determine the thickness of the gel and therefore the quantity of DNA that can be loaded The quality of product separation is affected by

Pouring a gel with molten agarose that is too hot will distort the casting tray. Consult manufacturer's recommendations, and when the solution is cool enough, pour the gel and remove any bubbles with a pipette tip. Leave to set

the polymerization process. Factors influencing polymerization are listed in *Table 11*

Mix the acrylamide stock to the desired concentration (v/v) with 1 × electrophoresis buffer

For denaturing PAGE, add a suitable denaturing agent at this stage (e.g. urea, 7 M final concentration). Do not use alkali, which deaminates acrylamide, or methylmercuric acid, which inhibits polymerization

Add a suitable volume of TEMED and APS to initiate polymerization (see *Table 11*)

Using a syringe or pipette, pour the acrylamide solution between sealed plates. Insert the comb immediately and leave to set

---

Abbreviations: APS, ammonium persulfate; TEMED, *N,N,N',N'*-tetramethylethylenediamine; v/v, volume/volume; w/v, weight/volume.

Table 8. Resolution of DNA in agarose

Agarose concentration (% w/v)	Effective separation range (kbp linear dsDNA)
2.0	0.1–3
1.5	0.2–4
1.2	0.4–6
1.0	0.5–7
0.8	0.7–10
0.6	1–20
0.3	5–60

Table 9. Resolution of DNA in polyacrylamide

Percentage of acrylamide (%w/v)	Effective separation range (bp linear dsDNA)
20.0	5–100
15.0	20–150
12.0	40–200
10.0	50–300
8.0	60–400
6.0	70–500
4.0	500–1500
3.0	1000–2500

Table 10. Types and applications of PAGE

Type of PAGE	Properties affecting mobility	Applications
Nondenaturing	Size Charge (pH dependent) Intramolecular secondary structure	Separation and purification of double-stranded fragments SSCP analysis
Denaturing	Size Charge (pH dependent)	Separation and purification of single-stranded fragments

Abbreviations: SSCP, single-strand conformation polymorphism.

Table 11. Factors affecting the polymerization of polyacrylamide gels

Factor affecting polymerization	Recommendations
Purity of acrylamide pH changes generated by the formation of acrylic acid during storage will alter fragment mobility. ionic contaminants accelerate or delay polymerization. Artifacts may be generated	Always use DNA grade acrylamide. Prepare fresh solutions every 8–10 weeks
Initiation of polymerization if polymerization occurs too quickly or too slowly, gel elasticity is decreased and turbidity is increased excess TEMED and APS causes a rapid shift in pH. This will accelerate the polymerization reaction	Add 35 µl TEMED and 700 µl 10% APS per 100 ml acrylamide solution
Reaction temperature polymerization is exothermic; this accelerates the reaction temperatures less than 20°C decrease gel elasticity and transparency	Polymerize at room temperature
Oxygen concentration the presence of excess oxygen inhibits the polymerization process	Degas all solutions at room temperature. Horizontal gels may be poured under nitrogen

Abbreviations: APS, ammonium persulfate; TEMED, *N,N,N',N'*-tetramethylethylenediamine.

Table 12. Parallel and perpendicular denaturing gradient gel electrophoresis (DGGE)

Type of DGGE	Direction of denaturing gradient	Technique	Application	Identification of polymorphisms
Perpendicular	Linear (increasing from left to right) At low concentrations, migration rate is dependent on molecular weight At high concentrations, the mobility of partially denatured molecules is less than the mobility of double-stranded molecules	Two PCR products are digested with a frequently cutting restriction endonuclease Fragments are electrophoresed together from a single well	Identification of polymorphisms in DNA from a single sample (e.g. from a single member of a pedigree) Establishing optimal gradients for parallel DGGE	Left or right shift in the melting curve of one of the fragments
Parallel	Linear (increasing from top to bottom)	PCR products are digested with frequently cutting restriction endonucleases Fragments are run from separate wells on two gels with overlapping concentration gradients	Searching for polymorphisms in multiple samples (e.g. members of the same pedigree) Two enzymes and two gels can detect up to 80% of possible base changes	Appearance of bands unique to certain samples Altered mobility of band(s) from at least one sample

Table 13. Methods and applications of DNA blotting

Application	Description	Function
Southern blot	Amplicons are separated by gel electrophoresis, denatured, neutralized, then transferred from the gel matrix to a solid support. The relative positions of separated fragments are conserved during transfer	Confirmation of PCR specificity
		Confirmation of product identity
	Any DNA is irreversibly bound to the membrane ready for hybridization with specific labeled probes	Detection of fragments not visible by gel electrophoresis and EtBr staining
Dot blot	Aliquots of the post-PCR are denatured, neutralized, then spotted on to a solid support in an ordered matrix using a blotting manifold	Allelic differentiation
	The DNA is irreversibly bound to the membrane ready for hybridization with specific labeled probes	Pathogen characterization
Reverse dot blot	Several specific oligonucleotide probes are immobilized on a solid support in an ordered matrix using a blotting manifold	Simultaneous analysis of multi-allelic variation
	The probes are hybridized with the denatured labeled products of a PCR in which several primer pairs may have been multiplexed	Routine high throughput typing

Detection of Amplified Products

Table 14. Properties of commercially available nucleic acid blotting membranes

Membrane type	Charge	Membrane properties
Biodyne A (GIBCO-BRL)	Neutral	Pure nylon
Neutral (Appligene)	Neutral	Pure nylon
Biodyne B (GIBCO-BRL, Promega)	Positive	Pure nylon
Positive (Appligene)	Positive	Pure nylon
Duralon-UV membranes (Stratagene)	Neutral	Supported nylon binding capacity >500 µg DNA cm^{-2}
Hybond-N (Amersham)	Neutral	Supported nylon binding capacity 480–600 µg DNA cm^{-2}
Zeta-Probe and Zeta-Probe Genomic Tested Membranes (Bio-Rad)	Positive	Supported nylon binding capacity 150 µg DNA cm^{-2}
Hybond-N+ (Amersham)	Positive	Supported nylon binding capacity 480–600 µg DNA cm^{-2}
Nitrocellulose membranes (Stratagene)	Neutral	Pure nitrocellulose binding capacity up to 80 µg DNA cm^{-2}
Duralose-UV membranes (Stratagene)	Neutral	Reinforced nitrocellulose binding capacity 80–100 µg DNA cm^{-2}

Reinforced membranes have increased durability and are recommended for sequential hybridizations. Note that preparation and immobilization conditions are specific for each membrane.

Table 15. Blotting procedures

	Southern blot	Dot blot	Reverse dot blot
Controls	Electrophorese positive and negative amplification controls alongside the samples under investigation	Positive amplification control using amplimers specific for a nonvariable sequence Positive and negative hybridization controls Hybridize membranes to control oligonucleotides as well as to labeled typing oligonucleotides	Positive amplification control using oligonucleotides specific for a nonvariable sequence Positive and negative hybridization controls Hybridize membranes to positive and negative control PCRs as well as to labeled PCR products
Procedure	Blot immediately after electrophoresis is complete; gel dehydration will decrease the efficiency of transfer Depurinate DNA fragments by soaking the gel in 0.25 M HCl for 15 min. Denature for 30 min in 0.5 M NaOH/1.5 M NaCl, then neutralize in 0.5 M Tris (pH 8.0)/1.5 M NaCl for 30 min. Agitate gently throughout all treatments Rinse the gel for 5 min in 10 × SSC. Transfer the DNA to the membrane according to one of the methods described in *Table 16*	Membrane preparation, loading volumes and blotting procedure should be according to manufacturers' recommendations Dilute each sample if required in 15 × SSC or 1 mM EDTA/20 mM Tris–HCl, pH 7.6 Denature each sample in an equal volume of 0.4 M NaOH/25 mM EDTA for 10 min at room temperature. Alternatively, heat denature at 95–100°C for 5–10 min Prewet the membrane in 2 × SSPE	Membrane preparation, loading volumes and blotting procedure should be according to manufacturers' recommendations Prepare one membrane per PCR. Spot an identical array of probes on to each membrane using the blotting manifold (e.g. Bio-Dot, Bio-Rad) Bind the oligonucleotides to the membrane as described in ref. 28 or 29 The blot is now ready for hybridization with labeled PCR products (see Section 4)

Continued

Table 15. Blotting procedures, *continued*

Southern blot	Dot blot	Reverse dot blot
Immediately after transfer, label the membrane and fix the DNA according to manufacturers' recommendations. The blot is now ready for hybridization	and assemble the blotting apparatus (e.g. Bio-Dot, Bio-Rad). Prepare one membrane per probe, and spot identical samples on to each Begin filtration before sample loading and continue gently until each sample is completely blotted Immediately after blotting, fix the DNA and remove any unbound nucleic acids by briefly rinsing in 2 × SSPE The blot is now ready for hybridization	

Solutions: 20 × SSC: 3 M NaCl, 0.3 M Na-citrate, pH 7.0; 20 × SSPE: 3 M NaCl, 200 mM NaH$_2$PO$_4$, 20 mM EDTA, pH 7.4.

Table 16. Methods of DNA transfer for Southern blotting

Method of transfer	Protocol	Comments
Capillary (Southern transfer) (e.g. Horizon Blot Transfer System, GIBCO-BRL)	Lay the treated gel on top of the blotting stage and fill the assembly tank with blotting buffer Cover the gel with the membrane and 2–4 sheets of Whatman 3MM paper, both soaked in 20 × SSC. Overlay with paper towels and a weighted glass plate	Nucleic acids will transfer in 1–24 h depending on size (1–20 kb, respectively) When assembling the blot, ensure that no air bubbles become trapped between the layers. Rolling a plastic pipette over each layer will help to ensure even transfer Do not let the paper towels touch the membrane or a

	Leave to transfer overnight	'short circuit' of buffer will be created. Sealing the edges of the membrane with Parafilm or Nescofilm will protect the blot
Vacuum (e.g. VacuGene XL Vacuum Blotting System, Pharmacia)	Place the treated gel on the membrane and assemble the vacuum chamber Apply a vacuum to draw the blotting buffer from the upper to the lower reservoir. Flow through the porous gel support will transfer the PCR fragments on to the membrane	Faster and more efficient than capillary transfer Apply the vacuum gradually and, if possible, maintain at a gentle pull; compressed gel has the capacity to retain DNA The pull of the vacuum must be uniform across the gel to ensure even fragment transfer
Electrophoresis (e.g. Vertical Gel system, Stratagene)	Soak the treated gel in 1 × TBE (see *Table 5*). Arrange the gel and membrane in the blotting tank and fill with 1 × TBE Apply a current to draw the DNA fragments through the gel on to the membrane	Transfer time is dependent on gel porosity, fragment size and electric field. 3 h is normally sufficient, even for larger fragments Buffer replacement or recirculation may be necessary to prevent electrophoretic exhaustion Electrophoretic blotting systems have a tendency to overheat. If the apparatus has no cooling core, perform the transfer in a cold lab
Semi-dry electrophoresis (e.g. Trans-Blot SD DNA/RNA Blotting Kit, Bio-Rad)	Soak the treated gel and filter papers in 1 × TBE (see *Table 5*). Assemble the transfer cell Apply a current. The saturated filter papers will act as buffer reservoirs to transfer the DNA fragments on to the membrane	Fast transfer time (PCR products will blot in <20 min) Low voltage means uniform temperature and therefore homogeneous transfer Minimal amounts of reagents required

Manufacturers' instructions should be adhered to when assembling apparatus.
Solutions: 20 × SSC: 3 M NaCl, 0.3 M Na-citrate, pH 7.0.

Detection of Amplified Products

Chapter 12 PURIFICATION OF PCR PRODUCTS – C.R. Newton

Products of the PCR will often require purification if detection is not the primary objective. The objective of any purification procedure is to remove other reaction components such as protein (e.g. bovine serum albumin (BSA), DNA polymerase), residual dNTPs and amplimers, primer dimers and other nonspecific amplification products. Size selection from a heterogeneous product population may be required. Amplicon purification is usually a prerequisite prior to:

1. Direct DNA sequencing;
2. Molecular cloning;
3. Probe preparation;
4. Restriction digestion (see Chapter 14, *Table 1* for restriction enzymes that *will* perform efficiently without amplicon purification).

ethanol precipitation. These procedures are also adequately documented [2]. Purifications may also be achieved using:

1. A variety of proprietary kits that are ideal for DNA isolation from agarose gels (see *Table 1*), usually these kits may also be applied to purification of PCR products directly from the reaction mix but care should be exercised as nonspecific amplification products will co-purify.
2. Ultrafiltration spin cartridges may also be used for purification from the reaction mixture, this procedure relies on repeated dilution and concentration of the reaction mix, low molecular weight compounds pass through a membrane during the concentration step. Molecular weight cut-offs are shown in *Table 2*.
3. Magnetic separations using streptavidin-coated magnetic beads (Dynabeads®, Dynal (UK) Ltd; Dynal Inc.; or

In many instances the success of the PCR will have been checked by gel electrophoresis, EtBr staining and UV irradiation (see Chapter 11). Subsequent purification from an excised gel band will therefore give the added benefit of an initial size selection and will achieve the removal of the unwanted reaction components discussed above. However, electrophoresis buffer, agarose, acrylamide and EtBr may replace the original unwanted components of the PCR. Physical purification methods for nucleic acids from gel matrices, such as electroelution, 'freeze squeeze' and the melting of low melting point (LMP) agar, are documented extensively elsewhere [1, 2]. It is recommended that these procedures are followed by butanol extraction (if the sample requires concentrating), phenol/chloroform extraction and

MagneSphere®, Promega) where a biotinylated amplimer has been employed (see Chapter 7) or a biotin dNTP incorporated. This is particularly applicable to purification of PCR products directly from the reaction mix, but care should be exercised as nonspecific amplification products and unincorporated amplimers will co-purify. The main utility here is in downstream detection, either with a specific probe or a binding partner specific for the other amplimer (see Chapter 11).

4. An alternative magnetic separation (Magpie™, NBL Gene Sciences) uses beads carrying ion-exchange groups that remove amplimers; dNTPs and DNA polymerase are subsequently removed by a precipitation step.
5. HPLC in rare situations.

Table 1. Kits for DNA isolation from solution or agarose gel

Supplier	Extraction principle	Comments	Product
Amicon	Ultrafiltration spin cartridge[a]	For recovery from solution [3]	Centricon®, Microcon™
Bio 101	Glass particle suspension	For recovery from solution or from agarose gel	Geneclean® (for TAE gels), Geneclean II® for TAE or TBE gels (see Chapter 11)
Bio Rad	Silica matrix, chaotropic salt	For recovery from solution or from agarose gel	Prep-A-Gene®
Clontech	Gel filtration spin column	For recovery from solution	CHROMA SPIN™-100, 200 or 400[b]
	Proprietary single-stranded affinity matrix	For separation and recovery of ds amplicons from ss amplicons and/or vice versa	SSAM™
DuPont NEN	Silica matrix, chaotropic salt	For recovery from solution	NENSORB®
5' to 3'	Gel filtration spin column	For recovery from solution containing up to 10 μg DNA	PCR SELECT-II®[c]
FMC BioProducts	Silica matrix, chaotropic salt, spin cartridge	For recovery from solution or from agarose gel	Spinbind™
GIBCO-BRL	Silica matrix	For recovery from solution or from agarose gel	GlassMAX™
ISS	Ultrafiltration spin cartridge[d]	For recovery from solution	UF30 Spin Filter®
Pharmacia	Gel filtration spin column	For recovery from solution	MicroSpin S-400 HR™
	Glass particle suspension	For recovery from agarose gel	Sephaglas™ Band Prep
Promega	Proprietry resin/vacuum-assisted minicolumn	For recovery from solution or from low-melt agarose gel	Wizard™ PCR prep.

Qiagen	Silica matrix, chaotropic salt	For recovery from agarose gel
	Anion exchange resin minicolumn	For recovery from solution
		QIAquick™ gel extraction kit
		QIAquick™ PCR purification kit or QIAGEN® PCR purification spin kit
Sigma	Glass particle suspension	Suitable for recovery from agarose or solution
		NucleiClean™
Stratagene	Proprietry matrix/syringe-assisted minicolumn	For recovery from solution
		PrimErase® Quik™

[a]Microcon™ microconcentrators or Centricon® concentrators 100 000 mol. wt membrane cut-off; for specific cut-off values see, *Table 2*.
[b]CHROMA SPIN™ 100 (amplicon > 140 bp, amplimer and short extension products < 30 bp), CHROMA SPIN™ 200 (amplicon > 150 bp, amplimer < 50 nt and short extension products < 45 bp), CHROMA SPIN™ 400 (amplicon > 600 bp, amplimer and short extension products < 100 bp).
[c]PCR SELECT-I® removes primers and recovers products > 20 bp, PCR SELECT-II® removes primers and primer-dimers and recovers products > 100 bp, PCR SELECT-III® removes primers and primer-dimers and recovers products > 300 bp.
[d]ISS spin filters, 30 000 mol. wt membrane cut-off.

Table 2. Molecular weight cut-offs for DNA purification by ultrafiltration

Membrane mol. wt cut-off	Nucleotide cut-off single-stranded	Nucleotide cut-off double-stranded
3000	10	10
10 000	30	20
30 000	60	50
50 000[a]	60	50
100 000	300	125

[a]Centricon® device only.

Chapter 13 CLONING PCR PRODUCTS — C.R. Newton

There is a variety of methods for cloning PCR products, including:

1. Use of TA vectors that provide single 3'T residues [1] that complement the single 3'A additions to amplimers produced by some thermostable DNA polymerases [2];
2. Directional cloning into vectors after having introduced specific restriction enzymes' recognition sites into the amplicon by incorporating the enzymes' recognition sites into the amplimers;
3. Introduction of an amplicon into a vector by insertion mutagenesis;
4. Blunt-end cloning of an amplicon into a vector;
5. The generation of restriction enzyme half-sites into the amplicon ends by incorporating them into the PCR primers;

```
                                          T7 transcription start
     5'...TGTAA TACGA CTCAC TATAG GGCGA ATTGG CCCCG ACGTC CTCCC GGCCG
     3'...ACATT ATGCT GAGTG ATATC CCGCT TAACC CGGGC TGCAG GAGGG CCGGC
           T7 promoter                        Apal    AatII    SphI

     CCATG GCCGC GGGATT3'                     ATCAC TAGTG CGGCC GCCTG CAGGT CGACC ATATG
     GGTAC CGGCG CCCTA  (cloned insert)    3' TTAGTG ATCAC GCCGG CGGAC GTCCA GCTGG TATAC
      NcoI   SacII                             Spel    NotI    PstI   SalI   NdeI

                               SP6 transcription start
                                                                                  LacZ
     GGAGA GCTCC CAACG CGTTG GATGC ATAGC TTGAG TATTC TATAG TGTCA CCTAA AT...3'
     CCTCT CGAGG GTTGC GCAAC GTACG TATCG AACTC ATAAG ACAGT GGATT TA...5'
      SacI       BstXI        NsiI                      SP6 promoter
```

Figure 1. pGEM-T vector (Promega) promoter and multiple cloning sites.

After preparing DNA from transformed organisms (Chapter 4), recombinant clones derived using any of the methods can be quickly screened for the presence of inserts by using the

6. Ligation-independent cloning using either T4 DNA polymerase or exonuclease III;
7. Incorporation of dUTP into PCR primers followed by UDG digestion and directional cloning.

Note

Although TA vectors have been developed specifically for PCR products, the 'Single dA' tailing kit (Novagen) allows the conversion of other DNA fragments, carrying any type of end, to ones having single 3' dA residues for cloning into these vectors.

Kits are available for each of the more generic systems (*Tables 1* and *2*) using vectors with a variety of restriction enzyme recognition sites flanking the cloning site. These are useful for subsequent manipulations of the cloned DNA and are shown in *Figures 1–15*. Where specific cloning junctions are desired, the researcher must employ one of the other cloning methods. *Table 3* outlines the benefits and drawbacks of each method.

primers that originally generated the amplicon or vector-specific primers. If it is necessary to determine the orientation of the amplicon within the vector, this may be accomplished using an amplicon-specific primer in conjunction with a vector-specific primer.

```
                                    T7 transcription start
5'....TTGTA ATACG ACTCA CTATA GGGCG AATTG GGCCC TCTAG ATGCA TGCTC GAGCG
3'....AACAT TATGC TGAGT GATAT CCCGC TTAAC CCGGG AGATC AACGT ACGAG CTCGC
           T7 promoter              ApaI  XbaI  NsiI  XhoI

GCCGC CAGTG TGATG GATAT CTGCA GAATT CGGCTT3'          AGCCG AATTC CAGCA
CGGCG GTCAC ACTAC CTATA GACGT CTTAA GCCGA  (cloned insert) 3' TTCGGC TTAAG GTCGT
 NotI  BstXI EcoRV        EcoRI                                 EcoRI

CACTG GCGGC CGTTA CTAGT GGATC CGAGC TCGGT ACCAA GCTTG ATGCA TAGCT TGAGT ATTCT
GTGAC CGCCG GCAAT GATCA CCTAG GCTCG AGCCA TGGTT CGAAC TACGT ATCGA ACTCA TAAGA
BstXI       SpeI  BamHI  SacI  KpnI   HindIII        NsiI
                                                           SP6 transcription start

ATAGT GTCAT CTAAA TA...3'
TATCA CAGTA GATTT AT...5'
SP6 promoter
```

Figure 2. pCR™II vector (Invitrogen) promoter and multiple cloning sites.

Cloning PCR Products

133

```
                                 T7 transcription start

5'... AATTA ATACG ACTCA GGGAG ACCCA AGCTT GGTAC CGAGC TCCGA TCCAC TAGTA
3'... TTAAT TATGC TGAGT GATAT CCCTC TGGGT TCGAA CCATG GCTCG AGCCT AGGTG ATCAT

                    T7 promoter        HindIII  KpnI  SacI  BamHI  SpeI

ACGGC CGCCA GTGTG CTGGA ATTCG GCTT3'                   AGCCG AATTC TGCAG ATATC
TGCCG GCGGT CACAC GACCT TAAGC CGA   (cloned insert) 3' TTCGG CTTAAG ACGTC TATAG

    EagI  BstXI        EcoRI                                   EcoRI  PstI  EcoRV

                                       SP6 transcription start

CATCA CACTG GCGGC CGCTC GAGCA TGCAT CTAGA GGGCC CTATT CTATA GTGTC ACCTA AACCT 3'
GTAGT GTGAC CGCCG GCGAG CTCGT ACGTA GATCT CCCGG GATAA GATAT CACAG TGGAT TCGA 5'

  BstXI   NotI  XhoI SphI/NsiI XbaI  ApaI                              SP6 promoter
```

Figure 3. pCR™ III eukaryotic expression vector (Invitrogen) promoter and multiple cloning sites.

```
                                                T7 transcription start

                                      5'...CTCTA ATACG ACTCA CTATA GGGAA AGCTT GCATG CCTGC AGGTC GACTC TAGAG
                                      3'...GAGAT TATGC TGAGT GATAT CCCTT TCGAA CGTAC GGACG TCCAG CTGAG ATCTC

LacZ →                                               T7 promoter       HindIII  SphI  PstI   SalI  XbaI

GATCT ACTAG TCATA TGGATT3'                         ATCGG ATCCC CGGGT ACCGA GCTCG AATTC
CTAGA TGATC AGTAT ACCTAA    (cloned insert) 3'     TAGCC TAGGG GCCCA TGGCT CGAGC TTAAG

     SpeI    NdeI                                   BamHI Smal KpnI SacI EcoRI
```

Figure 5. pT7 Blue vector (Novagen) promoter and multiple cloning sites.

```
ACTGG...3'
TGACC...5'
```

Figure 6. (below) pTOPE vector (Novagen) promoter and multiple cloning sites.

T7 transcription start

```
5'...TTGTA ATACG ACTCA CTATA GGGCG GAATT AATTC CGGTT ATTTT CCACC ATATT
3'...AACAT TATGC TGAGT GATAT CCCGC CTTAA TTAAG GCCAA TAAAA GGTGG TATAA
         T7 promoter                      ApaI        XbaI  NsiI        XhoI
```

```
GCCGT CTTTT CTTTG AAAAA CACGA
CGGCA GAAAA GAAAC TTTTT GTGCT
```

```
GGCAA....424bp....TCCAC GGGGA CGTGG TTTTG
CCGTT          AGGTG CCCCT GCACC AAAAC
```

```
                                                      ATCGG ATCCG AATTC
TGATA ATACC ATGGG CACCA CCCAT ATGGA TT3'  (cloned insert) 3' TTAGCC TAGG CTTAAG
ACTAT TATGG TACCC GTGGT GGGTA TACCT A
         MetAl aThrT hrHis MetAs p                                 BamHI    EcoRI
    NcoI  MscI         NdeI
              BstXI
```

```
GAGCT CCGTC GACAA GCTTG CGGCC GCATC CGAGC ACCAC CACCA CCACC ACTGA GATCT
CTCGA GGCAG CTGTT CGAAC GCCGG CGTAG GCTCG TGGTG GTGGT GGTGG TGACT CTAGA
 SacI  SalI       NotI         XhoI                                  BglII
              BstXI
```

```
CCGAT CCTCT AGAGT CGATC GACCT GCAGG CATGC A...3'
GGCTA GGAGA TCTCA GCTAG CTGGA CGTCC GTACG T...5'
  XbaI              PstI  SphI
```

T7 transcription start

```
5'....AATTA ATACG ACTCA CTATA GGGAG ACCAC AACGG TTTCC CTCTA GAAAT AATTT
3'....TTAAT TATGC TGAGT GATAT CCCTC TGGTG TTGCC AAAGG GAGAT CTTTA TTAAA
         T7 promoter                                              XbaI
```

```
                              ---rbs---                    -------T7·Tag-------
TGTTT AACTT TAAGA AGGAG ATATA CATAT GGCTA GCATG ACTGG TGGAC AGCAA ATGGG
ACAAA TTGAA ATTCT TCCTC TATAT GTATA CCGAT CGTAC TGACC ACCTG TCGTT TACCC
                              Me taIaS erMet ThrGl yGlyG lnGln MetGl
                                       NheI                             SacII
```

```
T....706bp...GAGTT TGTAG AAGTT CCGCA CCTCA CCGCT GGTGG TGCTG GTACC GCGGA
A....235aa...CTCAA ACATC TTCAA GGCGT GGAGT GGCGA CCACC ACGAC CATGG CGCCT
y            Gluva lValG lyLys ProHi sLeuT hrAla GlyGl yAlaG lyThr AlaAs
```

```
TTCCA GCTTG GTACC GAGCT CGATT GGATC CACTA GTAAC GGCCG CCAGT GTGCT GGAAT TTCTG
AAGTT CGAAC CATGG CTCGA GCTAA CCTAG GTGAT CATTG CCGGC GGTCA CACGA CCTTA AAGAC
pSerS erLeu ValPr oSerS erLeu ProIl eGlyS erThr SerAs nGlyA rgPro ValCy sGlyI lePheCy
   HindIII     SacI  BamHI   SpeI                         BstXI          EcoRI   PstI
```

```
              ATCCA TCACA CTGGC GGCCG CTCGA GCAGA TCCGG CTGCT AGCAG ACTAG
CAGATT3' (cloned insert) 3' TTAGGT AGTGT GACCG CCGGC GAGCT CGTCT AGGCC GACGA TCGTC TGATC
GTCTA
sArg
                                 BstXI    NotI    XhoI
```

```
AACAA AGCCC GAAAG GAAGC TGAGT TGGCT GCTGC CACCG CTGAG CAACA ACTAG
TTGTT TCGGG CTTTC CTTCG ACTCA ACCGA CGACG GTGGC GACTC GTTGT TGATC
uThrL ysPro GluAr gLysL euSer Trple uLeuP roPro LeuSe rAsnA snEnd
```

```
CATAA CCCCT TGGGG CCTCT AAACG GGTCT TGAGG GGTTT TTT G...3'
GTATT GGGGA ACCCC GGAGA TTTGC CCAGA ACTCC CCAAA AAA C...5'
           T7 terminator
```

Figure 4. pCITE vector (Novagen) promoter and multiple cloning sites.

```
LacZ
5'...TCTAA TACGA CTCAC TATAG GGAAA GCTTG CATGC CTGCA GGTCG ACTCT AGAGG
3'...AGATT ATGCT GAGTG ATATC CCTTT CGAAC GTACG GACGT CCAGC TGAGA TCTCC

      T7 promoter          HindIII SphI   PstI   SalI     XbaI

                          ATCGG ATCCC CGGGT ACCGA GCTCG AATTC
(cloned insert)  3' TTAGCC TAGGG GCCCA TGGCT CGAGC TTAAG

                          BamHI Smal KpnI  SacI   EcoRI

ATCTCA CTAGT CATAT GGATT3'
TAGAT GATCA GTATA CCTA...5'

 SpeI   NdeI

ACTGG CCGTC GTTTA CA...3'
TGACC GGCAG CAAAT GT...5'
```

Figure 7. pMOS*Blue* vector (Amersham) promoter and multiple cloning sites.

```
5'...CACCG CCAGT GAATT CGAGC TCGGT ACCCG GGGAT CCTCT AGAGT CGACC TGCAG GCCGC GGGCG CGCCG GGATC CTTAA GAATT
3'...GTGGC GGTCA CTTAA GCTCG AGCCA TGGGC CCCTA GGAGA TCTCA GCTGG ACGTC CGGCG CCCGC GCGGC CCTAG GAATT CTTAA

         EcoRI  SacI   KpnI  Smal           Ascl                BamHI         PacI

                                   5' CCT GATC CTCTA GAGTC GACTG
               (cloned insert)      3' GGA CCTAG GAGAT CTCAG CTGAC L

                                      BamHI  XbaI    SalI

TCTAG ACGTT TAAAC AGG3'                        ————> LacZ
AGATC TGCAA ATTTG TCC5'

 XbaI   PmeI

TTTAA ACTGG CAGGC ATGCA AGCTT GGCGT AAT...3'
AAATT TGACC GTCCG TACGT TCGAA CCGCA TTA...5'

 PmeI   PstI  SphI   HindIII
```

Figure 9. pNoTA vector (5' to 3' Inc.) multiple cloning sites.

```
5'...CTA TAAAG ATAGG CGTAT CACGA GGCCC TTTCG TCTTC AAGAA TTGAT CATAC GAATT
3'...GAT ATTTT TATCC GCATA GTGCT CCGGG AAAGC AGAAG TTCTT AACTA GTATG CTTAA

TTACC CTCGC TTCCA CGACA ACACC GATAA TCTTG CAGTT CCCGT TGATA GGAGT CATAG
AATGG GAGCG AAGGT GCTGT TGTGG CTATT AGAAC GTCAA GGGCA ACTAT CCTCA GTATC

GCCAT GAAGG ATTCA GG3'                  5' CCT TTCAG GTACT TCTGA CCGCC ATCTA
CGGTA CTTCC TAAGT CC5'  (cloned insert) 3' GGA AAGTC CATGA GAGCT GGCGG TAGAT

TGACC AGTTT CTTGA ATGTT GCTTC GTTCA GCATC GTCAG TTTGG CTACA ACAAG GCTTC
ACTGG TCAAA GAACT TACAA CGAAG CCAGT CAGTC AAACC GATGT TGTTC CGAAG

CATTC ACTGG CTCGC GTCCA GTATC TACTA ACAC...3'
GTAAG TGACC GAGCG CAGGT CATAG ATGAT TGTG...5'
```

Figure 10. pCR-TRAP vector (GenHunter). Sequencing (outer) and PCR (inner) primers italicized.

Figure 8. pCR Script™ vector (Stratagene) promoter and multiple cloning sites. The *Srf*I cloning site is indicated by bold type.

```
                                    T7 transcription start
                                           ↓
5'...CCCGT AATAC GACTC ACTAT AGGGC GAATT GGGTA CCGGG CCCCC CCTCG AGGTC GACGG
3'...GGCCA TTATG CTGAG TGATA TCCCG CTTAA CCCAT GGCCC GGGGG GGAGC TCCAG CTGCC
    BssHII               T7 promoter         KpnI  ApaI         XhoI   SalI

TATCG ATAAG CTTGA TATCG AATTC CTGCA GCCCG GGGGA TCCAC TAGTT CTAGA GCGGC CGCCA
ATAGC TATTC GAACT ATAGC TTAAG GACGT CGGGC CCCCT AGGTG ATCAA GATCT CGCCG GCGGT
ClaI  HindIII  EcoRV  EcoRI  PstI  SmaI  NsiI  SrfI/SmaI          NotI/EagI

                                                T3 transcription start
                                                          ↓
CCGCG GTGGA GCTCC AGCTT TTGTT CCCTT TAGTG AGGGT TAATT GCGCC CTTGG...3'
GGCGC CACCT CGAGG TCGAA AACAA GGGAA ATCAC TCCCA ATTAA CGCGG GAACC...5'
 SacII  SacI                                T3 promoter    BssHII
                                                                ──LacZ──→
```

Figure 11. pDIRECT vector (Clontech) promoter and multiple cloning sites.

```
                                 T7 transcription start
                                         ↓
5'...TTGT ATTGT AATAC GACTC ACTAT AGGGC GAATT GGAGC TCCAC CGCGG TGGCC GCCGC
3'...AACA TAACA TTATG CTGAG TGATA TCCCG CTTAA CCTCG AGGTG GCGCC ACCGG CGGCG
                         T7 promoter              SacI     SacII   NotI/EagI

                                             5' GGGGCCAGCCAGTTGAA TTCGA TATCA
TCTAG AGGAT CCAA                             AACTT AAGCT ATAGT
AGATC TCCTA GGTT GACCAAGCCCGG 5' (cloned insert)   EcoRI   EcoRV
 XbaI   BamHI                                            T3 transcription start
                                                                   ↓
AGCTT ATCGA TACCG TCGAC CTCGA GGGGG GGCCC CCGGG GGTAC CCAGC TTTTG TTCCC TTTAG
TCGAA TAGCT ATGGC AGCTG GAGCT CCCCC CCGGG GGCCC CCATG GGTCG AAAAC AAGGG AAATC
HindIII ClaI      SalI   XhoI              ApaI    KpnI

TGAGG GTTAA TT TCG...3'
ACTCC CAATT AA AGC...5'
                T3 promoter

                    ──LacZ──→
```

Cloning PCR Products

137

Figure 12. pAMP1 vector (GIBCO-BRL) promoter and multiple cloning sites.

```
LacZ ──→         T7 transcription start
       5'... TCTAA TACGA CTCAC TATAG GGAA AGCTG CCTGC GTACG CCTGC AGGTA CCGGT CCGGA ATTCC
       3'... AGATT ATGCT GAGTG ATATC CCTT TCGAC CATGC GGACG CATGC GGCCA TCCAT GGCCT TAAGG
                         T7 promoter                              PstI      SphI         EcoRI

                                                     GCG GCCGC TCTAG AGGAT CCAAG
                                    CGGGT CGACATCATCAT 3'                                         
                                    GCCCA GCT            (cloned insert) 3' ATCATCATCCGC CGGCG AGATC TCCTA GGTTC
                                    SmaI SalI                              NotI/XmaIII XbaI BamHI HindIII

                              SP6 transcription start
                                    ←──────
CTTAC GTACG CCGTGC ATCGC ACGTC TCTTC TATAG TGTCA CCTAA ATTC... 3'
GAATG CATGC GGCACG TACGC TGCAG AGAAG ATATC ACAGT GGATT TAAG... 5'
SnaBI MluI SphI AatII                            SP6 promoter
```

Figure 13. pAMP10 vector (GIBCO-BRL) promoter and multiple cloning sites.

```
LacZ ──→         T7 transcription start
       5'... TCTAA TACGA CTCAC TATAG GGAA AGCTG CCTGC ATCCG CAGT GGAAT CCAGC TCGGT ACCCG GGGAT
       3'... AGATT ATGCT GAGTG ATATC CCTT TCGAC GGACG TAGGC GTCA CCTTA GGTCG AGCCA TGGGC CCCTA
                         T7 promoter                           EcoRI    SstI        KpnI XmaI BamHI

                                                            GGG CCGGC GCTCT AGAGT CAACC TGCAG
                                    CCAACGCCGCTACAGT3'                                         
                                    GGT            (cloned insert) 3' TCTCTAGAGGATCC ACCGG CGAGA TCTCA GCTGG ACGTC
                                    MluI SpeI                         BglII NcoI NotI XbaI SalI PstI

                              SP6 transcription start
                                    ←──────
GCATG CAAGC TTGGG GTAAT CATGG TCATA GCTGT TTCCT GTG... 3'
CGTAC GTTCG AACCC CATTA GTACC AGTAT CGACA AGGA CAC... 5'
SphI HindIII
```

Figure 14. pAMP18 vector (GIBCO-BRL) multiple cloning sites.

```
       5'... AC GTTGT AAAAC GACGG CCAGT GCCAA GCTTG CATGC CTGCA GGTCG ACTCT AGAGC
       3'... TG CAACA TTTTG CTGCC GGTCA CGGTT CGAAC GTACG GACGT CCAGC TGAGA TCTCG
                                                 HindIII SphI PstI SalI XbaI

                                           GG ATCCC CGGGT ACCGA GCTCG
                                              TAGGG GCCCA TGGCT CGAGC
                                              BamHI XmaI KpnI SstI

CCCCGGG GGGCCG CCCATG TGATCATGCCAACC TAGGG GGCCCA TGGCT CGAGC
                  (cloned insert) 3' TGATCATGCCAACC TAGG...
                  NotI NcoI BglII
                  Spel MluI

AATTC GTAAT CATGG TCATA GCTGT TTCCT GTG... 3'
TTAAG CATTA GTACC AGTAT CGACA AAGGA CAC... 5'
EcoRI
```

Figure 15. pAMP19 vector (GIBCO-BRL) multiple cloning sites.

```
       5'... CC CAGTC ACGAC GTTGT AAAAC GACGG CCAGT GAATT CGAGC TCGGT ACCCG GGGAT
       3'... GG GTCAG TGCTG CAACA TTTTG CTGCC GGTCA CTTAA GCTCG AGCCA TGGGC CCCTA
                                                      EcoRI SstI KpnI XmaI BamHI

                                           GGG CCGGC GCTCT AGAGT CAACC TGCAG
                                           CCC GGCCG CGAGA TCTCA GCTGG ACGTC
                                           NotI XbaI SalI PstI

CCAACGCCGCTACAGT3'
GGT            (cloned insert) 3' TCTCTAGAGGATCC
MluI SpeI                         BglII NcoI

GCATG CAAGC TTGGG GTAAT CATGG TCATA GCTGT TTCCT GTG... 3'
CGTAC GTTCG AACCC CATTA GTACC AGTAT CGACA AGGA CAC... 5'
SphI HindIII
```

Table 1. Features of PCR cloning vectors: I

Vector	Supplier	Cloning method	Resistance markers	f1 Origin of replication[a]	Blue/white selection[b]	Sense transcripts[c]	Antisense transcripts[c]
pGEM-T	Promega	TA[1,2]	Amp	Yes	Yes	Yes	Yes
pCR II	Invitrogen	TA[1,2]	Amp + Kan	Yes	Yes	Yes	Yes
pCR III	Invitrogen	TA[1,2]	Amp + Kan + Neo	Yes	No	Yes	Yes
pCITE	Novagen	TA[1,2]	Amp	Yes	No	Yes	Yes
pT7 Blue	Novagen	TA[1,2]	Amp	Yes	Yes	Yes	No
pTOPE	Novagen	TA[1,2]	Amp	Yes	No	Yes	No
pMOS*Blue*	Amersham	TA[1,2]	Amp	Yes	Yes	Yes	No
pCR Script	Stratagene	Blunt-end ligation	Amp	Yes	Yes	Yes	Yes
pNoTA	5′ to 3′	Blunt-end ligation	Amp	No	Yes	No	No
pCR-TRAP	GenHunter	Blunt-end ligation	Tet	No	No	No	No
pDIRECT	Clontech	LIC [4–6]	Amp	Yes	Yes	Yes	Yes
pAMP 1	GIBCO-BRL	UDG [7,8]	Amp	Yes	Yes	Yes	Yes
pAMP 10	GIBCO-BRL	UDG [7,8]	Amp	Yes	Yes	Yes	Yes
pAMP 18	GIBCO-BRL	UDG [7,8]	Amp	No	Yes	No	No
pAMP 19	GIBCO-BRL	UDG [7,8]	Amp	No	Yes	No	No

Abbreviations: Amp, ampicillin resistance; Kan, kanamycin resistance; *LacZ*, β-galactosidase α-peptide; LIC, ligation independent cloning; Neo, neomycin phosphotransferase; Tet, tetracycline; UDG, uracil–DNA-glycosylase.

[a]For single-stranded DNA production (sense strand) for mutagenesis and dideoxy-sequencing.

[b]*LacZ* α-peptide of β-galactosidase complements *lac* deletion mutation in *E. coli* host strains supplied with the vectors having blue/white selection.

[c]*In vitro* transcription for generation of RNA probes, *in vitro* translation or subtracted cDNA libraries.

Table 2. Features of PCR cloning vectors: II

Vector	Supplier	Directional cloning	Post PCR modification of amplicon[a]	Binding sites for pUC/M13 primers[b]	Binding sites for promoter primers	Special features
pGEM-T	Promega	No	No	f and r	T7 and SP6	
pCR II	Invitrogen	No	No	f and r	T7 and SP6	
pCR III	Invitrogen	No	No	No	T7 and SP6	Eukaryotic expression
pCITE	Novagen	No	No	f and r	T7 and T3	*In vitro* translation
pT7 Blue	Novagen	No	No	f and r[d]	T7	
pTOPE	Novagen	No	No			Screening peptide domains
pMOS*Blue*	Amersham	No	No	r	T7	
pCR Script	Stratagene	No	End polishing[b]	f and r	T7 and T3	
pNoTA	5' to 3'	No	End polishing and phosphorylation[c]	f and r	No	
pCR-TRAP	GenHunter	No	No	[d]	[d]	Positive selection for recombinants[e]
pDIRECT	Clontech	Yes	T4 polase + dTTP	f and r	T7 and T3	Ligation not required
pAMP 1	GIBCO-BRL	Yes	UDG digestion	f and r	T7 and SP6	Ligation not required
pAMP 10	GIBCO-BRL	No	UDG digestion	f and r	T7 and SP6	Ligation not required: bacterial expression
pAMP 18	GIBCO-BRL	Yes	UDG digestion	f and r	No	Ligation not required
pAMP 19	GIBCO-BRL	Yes	UDG digestion	f and r	No	Ligation not required

Abbreviations: f, forward sequencing primer; *LacZ'*, β-galactosidase α-peptide; r, reverse sequencing primer; UDG, uracil–DNA-glycosylase.

[a] Assumes *Taq* DNA polymerase used for the PCR. For TA cloning it is essential to use a polymerase that generates 3'A addition to amplicons. End polishing will not be required for those enzymes that generate blunt ends (see Chapter 5, *Table 2*).

[b] Blunt-end ligation, whose efficiency is enhanced giving high yield and low background by being performed concurrently with a restriction enzyme digestion [3].

[c] The reagents for this step are provided in the kit.

[d] Alternative primers are available flanking the polylinker.
[e] A particularly useful feature of this vector is that the cloning site is within the gene encoding a repressor protein controlling the expression of tetracycline resistance, therefore only recombinant clones grow under tetracycline selection.

Table 3. Attributes of PCR cloning systems

Method	Useful attributes	Limitations	Availability in kit form	References
TA complementarity	• Simple procedure • No post PCR modification of amplicon required	• Can only be used for amplicons generated by polymerases that produce 3'A additions to product • Not directional • Cannot 'tailor' cloning junctions • High nonrecombinant background possible due to poor efficiency of T addition to vector or exonuclease removal of Ts from vector	• Variety of commercially available kits	1
Cloning after restriction enzyme digestion	• Can 'tailor' cloning junctions to some extent • Directional if two different restriction enzymes are used	• Larger amplimers required • Post PCR restriction digest(s) required (see Chapter 14, *Tables 1–3*)	No	9
Insertion mutagenesis	• Can completely 'tailor' cloning junctions • Directional	• Vector required that can generate single-stranded DNA	No	10
Blunt-end cloning	• Simple procedure • High efficiencies possible, particularly when concurrent restriction digestion is employed during ligation	• Requires phosphorylated amplimers or post PCR phosphorylation of amplicon unless concurrent restriction digestion is employed during ligation	• Yes, employs concurrent restriction digestion during ligation	3

Continued

Table 3. Attributes of PCR cloning systems, *continued*

Method	Useful attributes	Limitations	Availability in kit form	References
Restriction enzyme half sites	• Can 'tailor' cloning junctions to some extent • Directional	• Complex, multistep procedure • Requires phosphorylated amplimers or post PCR phosphorylation of amplicon • Amplicons derived using polymerases that generate 3'A additions to amplicon (see Chapter 5, *Table 2*) must be 'polished'	No	11
Ligation-independent cloning – T4 DNA polymerase treatment in the presence of a single dNTP	• Can 'tailor' cloning junctions to some extent • Very high efficiencies possible • Directional	• May require the vector to be PCR amplified unless using kit • Not recommended for amplicons generated using a DNA polymerase with 3'–5' exonuclease activity[a]	Yes	4–6
Ligation-independent cloning – UDG digestion of amplicons derived from dUTP-substituted primers	• Can 'tailor' cloning junctions to some extent • Very high efficiencies possible • Directional	• Requires the vector to be PCR amplified unless using kit • Requires 5' end dUTP substituted amplimers • Not recommended for amplicons generated using a DNA polymerase with 3'–5' exonuclease activity[a]	Yes	7,8

Chapter 14 MISCELLANEOUS DATA – C.R. Newton

1 Restriction endonuclease activity in PCR buffer – P. Eastlake, M. Pinkney and P. Gorringe

For many post-PCR enzymatic reactions it is considered necessary to remove components of amplification reactions such as dNTPs, amplimers, PCR buffer and *Taq* DNA polymerase (see Chapter 12). There are circumstances, however, when a simple diagnostic restriction digest is all that is required; in this case the enzyme activity needs to be known in PCR conditions. Restriction digestion pre-PCR can also be applied for the elimination of PCR product carry-over, see Chapter 9. *Table 1* shows the results of enzyme assays in PCR buffer and in completed PCR reactions. Using the percentage activity given, the quantity of enzyme required for restriction can be estimated. For some enzymes (recorded as 0–25% activity), restriction is not advised. For these enzymes, DNA should be purified before restriction since over-digestion under unfavorable conditions can lead to star activity giving non-specific bands. There are two main reasons why a particular enzyme does not perform well in PCR reactions:

1. pH – Most PCR buffers are in the range of 8.3–8.8 at 25°C, so that at 72°C the pH will be 7.0–7.5. For restriction enzymes used at 37°C, the pH will only drop to 8.0–8.5. Most restriction enzymes have optimum activity in the range of pH 7.0–8.0.
2. Magnesium concentration – most PCR buffers have a final Mg^{2+} concentration of 1.5 mM. The optima for restriction endonucleases varies greatly; some have a broad range over 1–15 mM, others have a distinct optimum around 5–8 mM.

In general, there is little difference between assays in PCR buffer and those performed in completed PCR reactions. This would suggest that dNTPs and primers do not have an effect on restriction.

A surprising result is that several enzymes perform better in PCR buffer than in their standard recommended buffer. *RsaI* and *XbaI* are the most notable examples. However, care should be taken not to incubate reactions for extended periods as elevated pH is one of the factors that will promote star activity.

Taq DNA polymerase has approximately 2.5% activity at 37°C and it should be noted that 3' recessed cohesive ends produced by restriction will act as a template for the polymerase [1]. Although this will not affect the restriction pattern, it will affect subsequent steps such as cloning.

It should also be noted that enzyme activity can vary according to the proximity of their recognition sites to the ends of PCR fragments (see *Tables 2* and *3*).

5 Misincorporation rates by *Taq* DNA polymerase

Table 14 shows relative likelihoods of nucleotide misincorporation by *Taq* DNA polymerase.

6 Metric conversions and other DNA conversion data

Tables 15 and *16*, respectively, give metric conversions and other miscellaneous DNA conversion data.

7 Commercially available kits and specialized reagents for PCR and associated techniques

PCR reagent kits (Applied Biosystems, GIBCO-BRL, Boehringer Mannheim)
PCR optimization kits (see Chapter 8)
ARMS kits (Kodak Clinical Diagnostics)
RT-PCR kits (Applied Biosystems, see *Table 13*; Stratagene)

2 Data for radioisotopes commonly used in PCR

Table 4 shows commonly used isotopes and their applications; *Tables 5, 6* and *7* give decay data for ^{32}P, ^{33}P and ^{35}S, respectively.

3 Genetic code, DNA degeneracy code

Tables 8 and *9* depict the genetic code; *Table 10* shows the DNA degeneracy codes.

4 Amino acid abbreviations and codon degeneracy and restriction site introduction via silent mutations

Table 11 shows amino acid abbreviations; *Table 12* depicts codon degeneracies; *Table 13* summarizes the possible amino acid sequences encoded by each reading frame comprising restriction enzyme recognition sequences.

Long-PCR kit (Applied Biosystems, Boehringer Mannheim)
cDNA synthesis kit (Boehringer Mannheim)
cDNA synthesis and cloning kit (Invitrogen)
Combined 5' and 3' RACE kit incorporating Long-PCR technology (Clontech)
Degenerate oligonucleotide primer PCR kit (Boehringer Mannheim)
5' RACE kit (Clontech, GIBCO-BRL)
3' RACE kit (GIBCO-BRL, Pharmacia, Stratagene)
In vitro expression of PCR products (GIBCO-BRL)
mRNA differential display kits (GenHunter, Display Systems)
PCR mimic construction kit (Clontech)
PCR carry-over prevention kit using UDG (Applied Biosystems)
PCR ELISA kit – digoxigenin labeling (Boehringer Mannheim)
PCR ELISA kit – digoxigenin detection (Boehringer Mannheim)
DNA immunoassay kits (see Chapter 11, *Table 2*) for cystic fibrosis and selected viral diseases (Sorin Biomedica)
Exon trapping kit (GIBCO-BRL)

Anti-*Taq* DNA polymerase antibody (TaqStart®, for hot start, Clontech)

Sequencing and cycle sequencing kits (Amersham, Applied Biosystems, Boehringer Mannheim, GIBCO-BRL, New England Biolabs, Pharmacia, Promega, Stratagene, see also Chapter 11)

Legionella detection kit, semi-quantitative (Applied Biosystems)

HLA forensic and typing kit (Applied Biosystems)

Forensic typing kit for six allelic markers (Applied Biosystems)

DNA and RNA purification kits (see Chapter 4)

Sex determination kit (Advanced Biotechnologies)

Site-directed mutagenesis kit (GIBCO-BRL, Takara Shuzo)

Cloning kits (see Chapter 13)

PCR product purification kits (see Chapter 12)

Mineral oil removal (OIL *AWAY*®, Bio 101)

Table 1. Restriction enzyme activity in PCR buffer

Enzyme	Reaction buffer[a]	Activity in PCR buffer[b]	Activity in completed PCR reaction[c]	Enzyme	Reaction buffer[a]	Activity in PCR buffer[b]	Activity in completed PCR reaction[c]
AluI	2	75–100	100	KpnI	3	75–100	75–100
ApaI	3	0–25	0–25	LspI	10	25–50	25–50
AvaI	4	75–100	75–100	MluI	6	25–50	25–50
BamHI	4	50–75	50–75	MspI	8	25–50	25–50
BclI	5	100	100	NcoI	8	75–100	ND
BglI	6	0–25	ND	NdeI	6	25–50	25–50
BglII	7	0–25	0–25	NruI	4	50–75	50–75
BsmI	5	0–25	0–25	NsiI	6	75–100	ND
BssHII	4	100	100	PstI	6	25–50	25–50
BstEII	4	100–125	ND	PvuII	5	100–125	>150
BstXI	6	0–25	ND	RsaI	8	>300	ND
CelII	6	0–25	ND	SacI	2	25–50	25–50
CfoI	8	125–150	75–100	SalI	6	25–50	ND
DdeI	6	50–75	ND	Sau3AI	2	50–75	25–50
DraI	5	75–100	75–100	ScaI	6	25–50	ND
EaeI	9	100	100	SinI	8	100	ND
EcoRI	6	0–25	75–100	SmaI	2	100	ND
EcoRV	4	75–100	ND	SphI	4*	50–75	25–50
HaeIII	5	75–100	75–100	SspI	6	50–75	ND
HincII	4	50–75	50–75	StuI	4	100	ND
HindIII	1	25–50	25–50	TaqI	4	100	ND
HinfI	6	25–50	25–50	XbaI	6	>200	>200
HpaI	2	75–100	ND	XhoI	6	25–50	ND

Continued

Table 1. Restriction enzyme activity in PCR buffer, *continued*

[a] Reaction buffers:
1. 10 mM Tris-HCl, pH 7.8, 50 mM KCl, 7 mM MgCl$_2$, 7 mM 2-mercaptoethanol, 100 µg ml^{-1} BSA.
2. 33 mM Tris-acetate, pH 8.2, 66 mM KAc, 10 mM MgAc, 0.5 mM DTT.
3. 6 mM Tris-HCl, pH 7.5, 6 mM NaCl, 6 mM MgCl$_2$, 6 mM 2-mercaptoethanol, 100 µg ml^{-1} BSA.
4. 10 mM Tris-HCl, pH 8.3, 100 mM NaCl, 5 mM MgCl$_2$, 1 mM 2-mercaptoethanol.
4*. As 4 with 100 µg ml^{-1} BSA.
5. 10 mM Tris-HCl, pH 7.8, 50 mM NaCl, 10 mM MgCl$_2$, 1 mM DTT.
6. 50 mM Tris-HCl, pH 7.8, 100 mM NaCl, 10 mM MgCl$_2$, 1 mM DTT.
7. 20 mM Glycine–NaOH, pH 9.5, 200 mM NaCl, 10 mM MgCl$_2$, 7 mM 2-mercaptoethanol.
8. 10 mM Tris-HCl, pH 7.8, 10 mM MgCl$_2$, 1 mM DTT.
9. 10 mM Tris-HCl, pH 7.5, 10 mM MgCl$_2$, 10 mM 2-mercaptoethanol, 100 µg ml^{-1} BSA.
10. 10 mM Tris-HCl, pH 7.5, 50 mM NaCl, 10 mM MgCl$_2$, 6 mM 2-mercaptoethanol, 100 µg ml^{-1} BSA, 0.1% (v/v) Triton X-100.

Activity in restriction buffer: Serial dilutions of enzyme were prepared and 5 µl added to 50 µl reaction mixture containing 1 µg λ DNA in single-strength restriction buffer. Reaction mixtures were incubated at the optimum temperature for each enzyme for 60 min. Reactions were halted by addition of 5 µl stop solution (0.25% (w/v) bromophenol blue, 40% (w/v) sucrose, 100 mM EDTA, pH 8.0, 1% (w/v) SDS), incubation at 65°C for 10 min and immediately electrophoresed on 0.5–1.0% (w/v) agarose gel. The lowest dilution giving a complete restriction pattern was used to calculate the original enzyme concentration.

[b] PCR buffer:

10 mM Tris-HCl, pH 8.3 @ 25°C, 50 mM KCl, 1.5 mM MgCl$_2$, 0.01% (w/v) gelatin.

Activity in PCR buffer; as for restriction buffer above, but single-strength PCR buffer was used. The value given in the table is the percentage activity compared to that obtained in the enzyme's optimum restriction buffer.

[c] Activity in PCR reaction: a Perkin-Elmer Lambda Control Reagents kit was used to generate PCR reactions. A 500 bp fragment of λ DNA was amplified under the following conditions: 1 ng λ DNA, 400 nM of each primer, 200 µM dNTPs, 1.25 units AmpliTaq in a volume of 50 µl in the PCR buffer given in note [b] above. Amplification used 25 cycles of 1 min at 94°C, 1 min at 37°C and 2 min at 72°C. Yield was approximately 1 µg/reaction. For enzyme assays, 25 µl PCR product was added to 1 µg λ DNA and diluted enzyme in 50 µl final volume.

Table 2. Cleavage close to the end of DNA fragments: I

Enzyme	Oligo sequence	Chain length	% Cleavage	
			2 h	20 h
AccI	**GGTCGACC**	8	0	0
	CG**GTCGAC**CG	10	0	0
	CCG**GTCGAC**CGG	12	0	0
AflIII	**CACATGTG**	8	0	0
	CC**ACATGT**GG	10	>90	>90
	CCC**ACATGT**GGG	12	>90	>90
AscI	**GGCGCGCC**	8	>90	>90
	A**GGCGCGCC**T	10	>90	>90
	TT**GGCGCGCC**AA	12	>90	>90
AvaI	**CCCCGGGG**	8	50	>90
	C**CCCGGGG**G	10	>90	>90
	TC**CCCGGGG**GA	12	>90	>90
BamHI	**CGGATCCG**	8	10	25
	CG**GGATCC**CG	10	>90	>90
	CGCG**GGATCC**GCG	12	>90	>90
BglII	**CAGATCTG**	8	0	0
	GA**AGATCT**TC	10	75	>90
	GGA**AGATCT**TCC	12	25	>90

Continued

Miscellaneous Data

Table 2. Cleavage close to the end of DNA fragments: I, *continued*

Enzyme	Oligo sequence	Chain length	% Cleavage 2 h	% Cleavage 20 h
BssHII	GG**GCGCGCC**	8	0	0
	AG**GCGCGCC**T	10	0	0
	TT**GGCGCGCC**AA	12	50	>90
BstEII	G**GGT(A/T)ACCC**	9	0	10
BstXI	AACTGCAGAA**CCAATGCATTGG**	22	0	0
	AAAACTGCAGAA**CCAATGCATTGG**AA	24	25	50
	CTGCAGAA**CCAATGCATTGG**ATGCAT	27	25	>90
ClaI	**CATCGAT**G	8	0	0
	G**ATCGATC**	8	0	0
	CC**ATCGAT**GG	10	>90	>90
	CCC**ATCGAT**GGG	12	50	50
EcoRI	GG**AATTC**C	8	>90	>90
	CGG**AATTC**CG	10	>90	>90
	CCGG**AATTC**CGG	12	>90	>90
HaeIII	GG**GGCC**CC	8	>90	>90
	AGC**GGCC**GCT	10	>90	>90
	TTGC**GGCC**GCAA	12	>90	>90

HindIII	**CAAGCTTG**	8	0	0
	CCAAGCTTGG	10	0	0
	CCCAAGCTTGGG	12	10	75
KpnI	**GGGTACCC**	8	0	0
	GGGGTACCCC	10	> 90	> 90
	CGGGGTACCCCG	12	> 90	> 90
MluI	**GACGCGTC**	8	0	0
	CGACGCGTCG	10	25	50
NcoI	**CCCATGGG**	8	0	0
	CATGCCATGGCATG	14	50	75
NdeI	CGCCATATGGCG	12	0	0
	GGGTTT**CATATG**AAACCC	18	0	0
	CGGAATT**CATATG**GAATTCC	20	75	> 90
	GGGAATTCCATATGGAATTCCC	22	75	> 90
NheI	**GGCTAGCC**	8	0	0
	CGGCTAGCCG	10	10	25
	CTAGCTAGCTAG	12	10	50
NotI	TTGCGGCCGCAA	12	0	0
	ATTT**GCGGCCGC**TTTA	16	10	10
	AAATAT**GCGGCCGC**TATAAA	20	10	10
	ATAAGAAT**GCGGCCGC**TAAACTAT	24	25	90
	AAGGAAAAAA**GCGGCCGC**AAAAGGAAA	28	25	> 90
Continued				

Miscellaneous Data

Table 2. Cleavage close to the end of DNA fragments: I, *continued*

Enzyme	Oligo sequence	Chain length	% Cleavage 2 h	% Cleavage 20 h
NsiI	TGCA**TGCATG**CA	12	10	>90
	CCAA**TGCAT**TGGTTCTGCAGT	22	>90	>90
PacI	**TTAATTAA**	8	0	0
	GT**TAATTAA**C	10	0	25
	CC**TTAATTAA**GG	12	0	>90
PmeI	**GTTTAAAC**	8	0	0
	G**GTTTAAAC**C	10	0	25
	GG**GTTTAAAC**CC	12	0	50
	AGCTT**GTTTAAAC**GGCGCGCCGG	24	75	>90
PstI	**GCTGCAGC**	8	0	0
	TGCA**CTGCAG**TCA	14	10	10
	AA**CTGCAG**AACCAATGCATTGG	22	>90	>90
	AAAA**CTGCAG**CCAATGCATTGGAA	24	>90	>90
	CTGCAGAACCAATGCATTGGATGCAT	26	0	0
PvuI	**CCGATCGG**	8	0	0
	AT**CGATCG**AT	10	10	25
	TC**GCGATCG**CGA	12	0	10
SacI	**CGAGCTCG**	8	10	10

SacII	**GCCGCGGC**	8	0	0
	TCCCCGCGGGGA	12	50	>90
Sall	**GTCGAC**GTCAAAAGGCCATAGCGGCCGC	28	0	0
	GCGTCGACGTCTTGGCCATAGCGGCCGCGG	30	10	50
	ACGCGTCGACGTCGGCCATAGCGGCCGCGGAA	32	10	75
Scal	**GAGTACTC**	8	10	25
	AAAGTACTTTT	12	75	75
Smal	**CCCGGG**	6	0	10
	CCCCGGGG	8	0	10
	CCCCCGGGGG	10	10	50
	TCCCCCGGGGGA	12	>90	>90
Spel	**GACTAGTC**	8	10	>90
	GG**ACTAGT**CC	10	10	>90
	CGG**ACTAGT**CCG	12	0	50
	CTAG**ACTAGT**CTAG	14	0	50
SphI	**GGCATGCC**	8	0	0
	CATG**CATGCA**TG	12	0	25
	ACATG**CATGCA**TGT	14	10	50
Stul	**AAGGCCTT**	8	>90	>90
	GA**AGGCCT**TC	10	>90	>90
	AAA**AGGCCT**TTT	12	>90	>90

Continued

Miscellaneous Data

Table 2. Cleavage close to the end of DNA fragments: I, *continued*

Enzyme	Oligo sequence	Chain length	% Cleavage 2 h	% Cleavage 20 h
XbaI	CT**CTAGAG**	8	0	0
	GC**TCTAGA**GC	10	>90	>90
	TGC**TCTAGA**GCA	12	75	>90
	CTAG**TCTAGA**CTAG	14	75	>90
XhoI	C**CTCGAG**G	8	0	0
	CC**CTCGAG**GG	10	10	25
	CCG**CTCGAG**CGG	12	10	75
XmaI	C**CCCGGG**G	8	0	0
	CC**CCCGGG**GG	10	25	75
	CCC**CCCGGG**GGG	12	50	>90
	TCCC**CCCGGG**GGGA	14	>90	>90

To test the ability of a restriction endonuclease to cleave a site which lies within a few bases of the end of a DNA fragment, a series of short, double-stranded oligonucleotides that contain the restriction endonuclease recognition site were digested. The table illustrates the varying requirements restriction endonucleases have for the number of bases flanking their recognition sequences (enzyme recognition sites appear in bold). This information may be helpful when designing extensions to amplimers for subsequent restriction enzyme digestion of the amplicon, choosing the order of addition of two restriction endonucleases for a double digest (a particular concern when cleaving sites close together in a polylinker), or when selecting enzymes most likely to cleave at the end of a DNA fragment. Data reproduced courtesy of New England Biolabs.

Table 4. Isotopes in PCR

Isotope	Compound	Application areas
^{32}P	[α-^{32}P]dATP	Amplicon labeling
^{32}P	[α-^{32}P]dCTP	Amplicon labeling
^{32}P	[α-^{32}P]dGTP	Amplicon labeling
^{32}P	[α-^{32}P]dTTP	Amplicon labeling
^{32}P	[γ-^{32}P]ATP	5'-end labeling of amplimers or amplicons
^{32}P	[α-^{32}P]ddATP	3'-end labeling of amplicons
^{33}P	[γ-^{33}P]ATP	DNA cycle sequencing
^{33}P	[α-^{33}P]dATP	Direct sequencing of amplicons
^{35}S	[^{35}S]dATPαS	Amplicon labeling and direct sequencing
^{35}S	[^{35}S]dCTPαS	Amplicon labeling and direct sequencing
^{35}S	[^{35}S]ATPγS	5'-end labeling of amplimers or amplicons

Refer to the safety note on p. xx regarding the use of ^{35}S-labeled compounds in thermocyclers.

Table 3. Cleavage close to the end of DNA fragments: II (from ref. 2)

Enzyme	Tail length (bp)	Digestibility
BamHI	3	Yes
EcoRI	4	Yes
HindIII	9	Yes
SalI	9	Yes
XhoI	10, 20	No, yes

Miscellaneous Data

Table 5. Decay table for ^{32}P

Days	Hours							
	0	12	24	36	48	60	72	84
0	1.000	0.976	0.953	0.930	0.908	0.886	0.865	0.844
4	0.824	0.804	0.785	0.766	0.748	0.730	0.712	0.695
8	0.679	0.662	0.646	0.631	0.616	0.601	0.587	0.573
12	0.559	0.546	0.533	0.520	0.507	0.495	0.483	0.472
16	0.460	0.449	0.439	0.428	0.418	0.408	0.398	0.389
20	0.379	0.370	0.361	0.353	0.344	0.336	0.328	0.320
24	0.312	0.305	0.298	0.291	0.284	0.277	0.270	0.264
28	0.257	0.251	0.245	0.239	0.234	0.228	0.223	0.217
32	0.212	0.207	0.202	0.197	0.192	0.188	0.183	0.179

Table 7. Decay table for ^{35}S

Days	Days						
	0	1	2	3	4	5	6
0	1.000	0.992	0.984	0.976	0.969	0.961	0.954
7	0.946	0.939	0.931	0.924	0.916	0.909	0.902
14	0.895	0.888	0.881	0.874	0.867	0.860	0.853
21	0.847	0.840	0.833	0.827	0.820	0.814	0.807
28	0.801	0.795	0.788	0.782	0.776	0.770	0.764
35	0.758	0.752	0.746	0.740	0.734	0.728	0.722
42	0.717	0.711	0.705	0.700	0.694	0.689	0.683
49	0.678	0.673	0.667	0.662	0.657	0.652	0.646
56	0.641	0.636	0.631	0.626	0.621	0.616	0.612
63	0.607	0.602	0.597	0.592	0.588	0.583	0.579
70	0.574	0.569	0.565	0.560	0.556	0.552	0.547
77	0.543	0.539	0.534	0.530	0.526	0.522	0.518
84	0.514	0.510	0.506	0.502	0.498	0.494	0.490

Refer to the safety note on p. xx regarding the use of ^{35}S-labeled compounds in thermocyclers.

Table 6. Decay table for ^{33}P

Days	0	1	2	3	4	5	6
0	1.000	0.973	0.947	0.921	0.897	0.872	0.849
7	0.826	0.804	0.782	0.761	0.741	0.724	0.701
14	0.683	0.664	0.646	0.629	0.612	0.595	0.579
21	0.564	0.549	0.534	0.520	0.506	0.492	0.479
28	0.466	0.453	0.441	0.429	0.418	0.406	0.395
35	0.385	0.374	0.364	0.355	0.345	0.336	0.327
42	0.318	0.309	0.301	0.293	0.285	0.277	0.270

Table 8. The genetic code

First position (5' end)	Second position				Third position (3' end)
	T	C	A	G	
T	Phe	Ser	Tyr	Cys	T
	Phe	Ser	Tyr	Cys	C
	Leu	Ser	Stop	Stop	A
	Leu	Ser	Stop	Trp	G
C	Leu	Pro	His	Arg	T
	Leu	Pro	His	Arg	C
	Leu	Pro	Gln	Arg	A
	Leu	Pro	Gln	Arg	G
A	Ile	Thr	Asn	Ser	T
	Ile	Thr	Asn	Ser	C
	Ile	Thr	Lys	Arg	A
	Met	Thr	Lys	Arg	G
G	Val	Ala	Asp	Gly	T
	Val	Ala	Asp	Gly	C
	Val	Ala	Glu	Gly	A
	Val	Ala	Glu	Gly	G

Miscellaneous Data

Table 9. Degeneracy codes

Symbol	Meaning	Complement
M	A or C	K
R	A or G	Y
W	A or T	W
S	C or G	S
Y	C or T	R
K	G or T	M
V	A or C or G	B
H	A or C or T	D
D	A or G or T	H
B	C or G or T	V
X/N	G or A or T or C	X/N

Table 11. Codon degeneracy

Number of codons	Amino acids
1	Met, Trp
2	Asn, Asp, Cys, Gln, Glu, His, Lys, Phe, Tyr
3	Ile
4	Ala, Gly, Pro, Thr, Val
6	Arg, Leu, Ser

Table 10. Amino acid abbreviations

Amino acid	Three-letter abbreviation	One-letter symbol
Alanine	Ala	A
Arginine	Arg	R
Asparagine	Asn	N
Aspartic acid	Asp	D
Cysteine	Cys	C
Glutamine	Gln	Q
Glutamic acid	Glu	E
Glycine	Gly	G
Histidine	His	H
Isoleucine	Ile	I
Leucine	Leu	L
Lysine	Lys	K
Methionine	Met	M
Phenylalanine	Phe	F
Proline	Pro	P
Serine	Ser	S
Threonine	Thr	T
Tryptophan	Trp	W
Tyrosine	Tyr	Y
Valine	Val	V

Table 12. Restriction site introduction via silent mutations

Enzyme	Recognition sequence	Amino acids, reading frame 1[a]			Amino acids, reading frame 2			Amino acids, reading frame 3[a]		
		1	2	3	1	2	3	1	2	3
AatII	GACGT/C	D	V	S	§RG	R	LPHQR	LS§WPQRMTKVAEG	T	S
Alw44I	G/TGCAC	V	H	T	CRSG	A	LPHQR	LS§WPQRMTKVAEG	C	T
ApaI	GGGCC/C	G	P	P	WRG	A	LPHQR	LS§WPQRMTKVAEG	G	P
BalI	TGG/CCA	W	P	HQ	LMV	A	IMTNKSR	FSYCLPHRITNVADG	G	HQ
BamHI	G/GATCC	G	S	P	WRG	I	LPHQR	LS§WPQRMTKVAEG	D	P
BclI	T/GATCA	§	S	HQ	LMV	I	IMTNKSR	FSYCLPHRITNVADG	D	HQ
BglII	A/GATCT	R	S	L	§QKE	—	FLSY§CW	LS§PQRITKVAEG	D	L
BssHII	G/CGCGC	A	R	A	CRSG	A	LPHQR	FSYCLPHRITNVADG	R	A
BspMII	T/CCGGA	S	G	DE	FLIV	R	IMTNKSR	FSYCLPHRITNVADG	P	DE
ClaI	AT/CGAT	I	D	IM	YHND	R	FLSY§CW	LS§PQRITKVAEG	S	IM
EcoRI	G/AATTC	E	F	S	§RG	—	LPHQR	LS§WPQRMTKVAEG	N	S
EcoRV	GAT/ATC	D	—	S	§RG	Y	LPHQR	LS§WPQRMTKVAEG	I	S
HindIII	A/AGCTT	K	L	FL	§QKE	A	FLSY§CW	LS§PQRITKVAEG	S	FL
HpaI	GTT/AAC	V	N	T	CRSG	§	LPHQR	LS§WPQRMTKVAEG	L	T
KpnI	GGTAC/C	G	T	P	WRG	Y	LPHQR	LS§WPQRMTKVAEG	V	P
MluI	A/CGCGT	T	R	V	YHND	A	FLSY§CW	LS§PQRITKVAEG	R	V
MscI	TGG/CCA	W	P	HQ	LMV	A	IMTNKSR	FSYCLPHRITNVADG	G	HQ
NaeI	GCC/GGC	A	G	A	CRSG	R	LPHQR	LS§WPQRMTKVAEG	P	A
NarI	GG/CGCC	G	A	P	WRG	R	LPHQR	LS§WPQRMTKVAEG	A	P
NcoI	C/CATGG	P	W	G	SPTA	M	VADEG	FSYCLPHRITNVADG	H	G
NdeI	CA/TATG	H	M	C§W	SPTA	Y	VADEG	FSYCLPHRITNVADG	I	C§W

Continued

Miscellaneous Data

Table 12. Restriction site introduction via silent mutations, *continued*

Enzyme	Recognition sequence	Amino acids, reading frame 1[a]			Amino acids, reading frame 2[a]			Amino acids, reading frame 3[a]		
		1	2	3	1	2	3	1	2	3
*Nhe*I	G/CTAGC	A	S	A	CRSG	§	LPHQR	LS§WPQRMTKVAEG	L	A
*Nru*I	TCG/CGA	S	R	S	FLIV	A	IMTNKSR	FSYCLPHRITNVADG	R	DE
*Nsi*I	ATGCA/T	M	Q	R	YHND	A	FLSY§CW	LS§PQRITKVAEG	C	IM
*Pst*I	CTGCA/G	L	Q	P	SPTA	A	VADEG	FSYCLPHRITNVADG	C	SR
*Pvu*I	CGAT/CG	R	S	A	SPTA	I	VADEG	FSYCLPHRITNVADG	D	R
*Pvu*II	CAG/CTG	Q	L	T	SPTA	A	VADEG	LS§WPQRMTKVAEG	S	C§W
*Sal*I	G/TCGAC	V	D	I	CRSG	R	LPHQR	LS§WPQRMTKVAEG	S	T
*Sca*I	AGT/ACT	S	T	L	§QKE	Y	FLST§CW	LS§PQRITKVAEG	V	L
*Sma*I	CCC/GGG	P	G	P	SPTA	R	VADEG	FSYCLPHRITNVADG	P	G
*Sna*BI	TAC/GTA	Y	V	H	LIV	R	IMTNKSR	FSYCLPHRITNVADG	T	Y§
*Spe*I	A/CTAGT	T	S	L	YHND	§	FLSY§CW	LS§PQRITKVAEG	L	V
*Sph*I	GCATG/C	A	C	A	CRSG	M	LPHQR	LS§WPQRMTKVAEG	H	A
*Sst*I	GAGCT/C	E	L	S	§RG	A	VADEG	FSYCLPHRITNVADG	S	S
*Sst*II	CCGC/GG	P	R	P	SPTA	A	VADEG	LS§WPQRMTKVAEG	R	G
*Stu*I	AGG/CCT	R	P	G	§QKE	A	FLSY§CW	LS§PQRITKVAEG	G	L
*Xba*I	T/CTAGA	S	R	L	FLIV	§	IMTNKSR	FSYCLPHRITNVADG	L	DE
*Xho*I	C/TCGAG	L	E	D	SPTA	R	VADEG	FSYCLPHRITNVADG	S	SR
*Xma*III	C/GGCCG	R	P	R	SPTA	A	VADEG	FSYCLPHRITNVADG	G	R

[a]Reading frames: XXX XXX, reading frame 1; XX XXX X, reading frame 2; X XXX XX, reading frame 3.
§Denotes termination;/shows restriction enzyme cleavage. Single-letter amino acid abbreviations are shown in *Tables 9* and *11*.

Table 13. *Taq* DNA polymerase misincorporation preference

CC				TC		AC		GT
AG	<	AA	<	CT	<<	CA	<<	TG
GA				TT				
GG								

Table 14. Metric conversions

1 g (gram)	= 1 g
1 mg (milligram)	= 10^{-3} g
1 μg (microgram)	= 10^{-6} g
1 ng (nanogram)	= 10^{-9} g
1 pg (picogram)	= 10^{-12} g
1 fg (femtogram)	= 10^{-15} g
1 ag (attogram)	= 10^{-18} g

Table 15. Other DNA data

1 μg ml^{-1} of DNA	= 3.08 μM phosphate
1 μg ml^{-1} of 1 kb of DNA	= 3.08 nM 5' ends
1 kb of DNA	= 6.5×10^5 Da of dsDNA (sodium salt)
1 kb of DNA	= 3.3×10^5 Da of ssDNA (sodium salt)
1 kb of DNA	= 3.4×10^5 Da of ssRNA (sodium salt)
Average molecular weight of a deoxynucleotide base	= 324.5 Da
Average molecular weight of a deoxynucleoside base pair	= 649 Da
Average molecular weight of a ribonucleotide base	= 340.5 Da

Miscellaneous Data

Table 16. RT-PCR kits

Attribute	GeneAmp EZ rTth RNA PCR kit	GeneAmp thermostable rTth reverse transcriptase RNA PCR kit	GeneAmp RNA PCR kit
Application	Screening cellular RNAs, RNA viruses	cDNA library construction, gene expression detection	cDNA library construction, gene expression detection
Template properties	General or with secondary structures	Secondary structures	General RNA without secondary structures
RT enzyme	rTth DNA polymerase	rTth DNA polymerase	MuLV reverse transcriptase
PCR enzyme	rTth DNA polymerase	rTth DNA polymerase	AmpliTaq DNA polymerase
RT temperature (°C)	60–70	60–70	42
Amplification temperatures (°C)	60/94	60/94	60/94
dUTP incorporation	+	+	−
AmpliWax compatible	−	−	+
Buffer	Mn^{2+}/Bicine	Mn^{2+}/Tris	Mg^{2+}/Tris
Average transcript length	1.3–3 kb	3 kb	3 kb
Primers recommended			
Specific	++	++	++
Oligo d(T)	+	+	+
Random hexamers	−	−	+
Positive control template and primers	+	+	+

Reproduced courtesy of Applied Biosystems.

Chapter 15 **TROUBLESHOOTING** – C.R. Newton

Table 1 lists the symptoms of the problems that may be experienced when practicing PCR, together with the possible causes and the requisite solutions to the problems.

Table 1. PCR troubleshooting guide

Symptom	Cause	Remedy
1. No or poor amplification	Incomplete denaturation of substrate DNA	Boil DNA in aqueous solution prior to combining with other reaction components
		Add 7-deaza-dGTP (3:1 to dGTP) in the reaction mixture
	Strong secondary structure in the template	Boil DNA in aqueous solution prior to combining with other reaction components
		Add DMSO at up to 10% and/or glycerol up to 12% and/or formamide up to 10% (determine exact amount(s) empirically)
		Use of N-terminal deletion of *Taq* DNA polymerase (see Chapter 5)
	Target sequence not present in sample	Carry out reaction with positive control

Continued

Table 1. PCR troubleshooting guide, continued

Symptom	Cause	Remedy
	Suboptimal thermal profile	Increase denaturation step to at least 1 min; allow 0.5 min kb^{-1} extension time (0.7 min kb^{-1} for long PCR), (ref. 1, see also Chapter 8) Check that annealing temperature is not too high, reduce appropriately
	Poor thermal conductivity from reaction block to reaction tube	Add a drop of oil to sample well of reaction block
	Incorrect primer sequence	Check amplimer sequence, especially at 3' end
	Degradation of primers when using 3' exo^{+} polymerases	Add enzyme or amplimers last and immediately prior to thermal cycling, preferably by hot start Synthesize amplimers with 3' phosphorothioate linkage Change DNA polymerase (see Chapter 5)
	Insufficient Mg^{2+} ions	Check buffer recipe: titrate magnesium over 1.0 mM to 10 mM range in 0.5 mM increments (see also Chapter 8)
	No vapor barrier when using thermal cycler without heated lid	Add mineral oil overlay or use AmpliWax PCR gems
	Nucleases in water	Autoclave water
	Nucleases in working/stock solutions	Replace working solutions first; if not remedied, replace stock solutions
	Nucleases from fingertips	Wear gloves

Cause	Solution
dNTPs are too acidic	dNTPs should be bought in buffered solution or alternatively dissolved in distilled water and the pH adjusted to pH 7.0 with sodium hydroxide
Inhibitors present in the sample	Dilute out inhibitors Purify/extract nucleic acid from the sample
Diethylpyrocarbonate (DEPC) present in the sample	Avoid the use of DEPC-treated water
Glove powder present in the sample	Use powder-free gloves Wash gloved hands prior to handling PCR tubes
Thermal cycler fault	Check performance of thermal cycler
Degraded, poor quality or incorrect sequence amplimers	Amplimer resynthesis
Combination of deoxyinosine containing amplimers with *Pfu* DNA polymerase	Use alternative DNA polymerase (Chapter 5) Use universal nucleoside (Chapter 7) to replace deoxyinosine
Incompatibility of the DNA polymerase with dUTP/UDG carry-over prevention	Check compatibility of the DNA polymerase (see Chapters 5 and 9); use alternative enzyme
EtBr omitted from gel and running buffer	Stain gel (see Chapter 11)
2. Amplification of only some samples in a batch	
Glycerol from enzyme stock solution causes enzyme to sink to bottom of master mix tube	Ensure that master mix is homogeneous by repeated aspiration by pipette tip prior to transfer of aliquots to reaction tubes
Thermal cycler fault	Check performance of thermal cycler
Glove powder present in some samples	Use powder-free gloves Wash gloved hands prior to handling PCR tubes

Continued

Table 1. PCR troubleshooting guide, continued

Symptom	Cause	Remedy
3. Smearing on gel electrophoresis	Extension time too long	Reduce extension to 1 min of 0.5 min kb^{-1} extension time (0.7 min kb^{-1} for long PCR), (see Chapter 8)
	Too much template	Titrate DNA (serial dilutions)
	Too much primer	Titrate amplimers (0.1 μM to 1.0 μM)
4. Ladders of PCR products	Nonspecific binding of amplimers to DNA	Try hot-start (see Chapter 8)
		Increase annealing temperature in 5°C increments
	Too much DNA polymerase added	Reduce DNA polymerase concentration
	Mg^{2+} too high	Check buffer recipe: titrate magnesium over 1.0 mM to 10 mM range in 0.5 mM increments
5. Primer-dimer formation	3' end complementarity of amplimers	Check for self-complementarity of individual amplimers at their 3' end
		Check for complementarity between amplimers at their 3' ends
	Annealing temperature too low	Increase annealing temperature in 5°C increments
		Check performance of thermal cycler
	Primer concentration too high	Titrate amplimers (0.1 μM to 1.0 μM)
6. Amplicon cannot be blunt-end cloned	Presence of nontemplate-derived single nucleotide addition at 3' end	Use alternative DNA polymerases (see Chapter 5)
		Polish amplicon using T4 or *Pfu* DNA polymerase [2]
7. Amplicon cannot be TA cloned	Absence of nontemplate-derived single nucleotide addition at 3' end	Use alternative DNA polymerase (see Chapter 5)

	3' T residues removed from vector by exonuclease	Use fresh vector preparation
8. Amplicon cannot be directionally cloned	Insufficient 5' extension to amplimers for efficient restriction enzyme digestion	Redesign amplimers (see Chapter 14) Concatemerize amplicon by blunt-end ligation prior to restriction enzyme digestion [3]
	Amplicon complexed with DNA polymerase	Deproteinize reaction mixture prior to restriction enzyme digestion (see Chapter 12)
9. Amplicon cannot be concatemerized and/or linkers cannot be ligated	Presence of nontemplate-derived single nucleotide addition at 3' end	Polish amplicon using T4 or *Pfu* DNA polymerase [2]
	Amplimers without 5' phosphate used	Phosphorylate amplimers or amplicon using T4 polynucleotide kinase and ATP
10. PCR products in −ve controls	PCR product carry-over	Refer to Chapter 9
11. Amplicon sequence errors	PCR-derived nucleotide misincorporation and misextension	Change DNA polymerase (see Chapter 5) Increase starting template Reduce number of thermal cycles
	dNTPs concentration too high	Decrease dNTPs concentration to below 200 μM

Chapter 16 MANUFACTURERS AND SUPPLIERS

Advanced Biotechnologies Ltd, Unit 7, Mole Business Park 3, Randalls Road, Leatherhead, Surrey KT22 7BA, UK.
Tel: 01372 360123. Fax: 01372 363263.

Amersham International plc, Amersham Place, Little Chalfont, Bucks HP7 9NA, UK.
Tel: 0800 616929. Fax: 0800 616927.
US address: 2636 South Clearbrook Drive, Arlington Heights, IL 60005, USA.
Tel: 708 593 6300. Fax: 708 593 8010.

Amicon, Upper Mill, Stonehouse, Gloucester, Glos GL10 2BJ, UK.
Tel: 01453 825181. Fax: 01453 826686.
US address: 72 Cherry Hill Drive, Danvers, MA 01923, USA.
Tel: 800 343 1397. Fax: 508 777 6204.

US address: 850 Lincoln Center Drive, Foster City, CA 94404, USA.
Tel: 415 570 6667. Fax: 415 572 2743.

Appligene, Pinetree Centre, Durham Road, Birtley, Chester-le-Street, Co. Durham DH3 2TD, UK.
Tel: 0191 492 0022. Fax: 0191 492 0617.
US address: 1177-C Quarry Lane, Pleasanton, CA 94566, USA.
Tel: 800 955 1274. Fax: 510 462 6247.

BIO 101, Inc, PO Box 2284, La Jolla, CA 92038-2284, USA.
Tel: 619 598 7299. Fax: 619 598 0116.
(UK distributor, Stratech Scientific Ltd.)

Amitof Biotech Inc., 14–20 Linden Street, Boston, MA 02134, USA.
Tel: 617 782 9242. Fax: 617 782 9352.

Amresco Inc., 30175 Solon Industrial Parkway, Solon, OH 44139, USA.
Tel: 216 349 2805. Fax: 216 349 1182.
(UK distributor, Camlab Ltd.)

AMS Biotechnology Ltd, 5 Thorney Leys Park, Witney, Oxon OX8 7GE, UK.
Tel: 01993 706500. Fax: 01993 706006.

Anachem Ltd, Anachem House, 20 Charles Street, Luton LU2 0EB, UK.
Tel: 01582 456666. Fax: 01582 391768.

Applied Biosystems Ltd (a division of Perkin-Elmer), Kelvin Close, Birchwood Science Park North, Warrington, Cheshire WA3 7PB, UK.
Tel: 01925 825650. Fax: 01925 828196.

Bio/Gene Ltd, Bio/Gene House, Greensbury Farm, Bolnhurst, Beds MK44 2ET, UK.
Tel: 01234 376762. Fax: 01234 378114.

Biometra Ltd, Whatman House, St Leonard's Road, 20/20 Maidstone, Kent ME 16 0LS, UK.
Tel: 01622 678872. Fax: 01622 752774.
US address: 550 North Reo Street, Tampa, FL 33609-1013, USA.
Tel: 813 287 5132. Fax: 813 287 5163.

Bioquote Ltd, The Raylor Centre, James Street, York, N. Yorks Y01 3DW, UK.
Tel: 01904 431402. Fax: 01904 431409.

Bio-Rad Laboratories Ltd, Bio-Rad House, Maylands Avenue, Hemel Hempstead, Herts HP2 7TD, UK.
Tel: 0800 181134. Fax: 01442 259118.
US address: Alfred Nobel Drive, Hercules, CA 94547, USA.
Tel: 510 741 1000. Fax: 510 741 1060.

Manufacturers and Suppliers

Bios Corporation, 291 Whitney Avenue, New Haven, CT, USA.
Tel: 800 678 9487. Fax: 203 562 9377.
(UK distributor, Scotlab Ltd.)

BioTherm Corporation, 9685-A Main Street, Fairfax, VA 22031, USA.
Tel: 703 425 1678. Fax: 703 425 1679.

Boehringer Mannheim (Diagnostics and Biochemicals) Ltd, Bell Lane, Lewes, East Sussex BN7 1LG, UK.
Tel: 01273 480444. Fax: 01273 480266.
US address: PO Box 50414, Indianapolis, IN 46250-0414, USA.
Tel: 317 849 9350. Fax: 317 576 2754.

Cambio Ltd, 34 Millington Road, Cambridge CB3 9HP, UK.
Tel: 01223 66500. Fax: 01223 350069.

Cruachem Ltd, Todd Campus, West of Scotland Science Park, Acre Road, Glasgow G20 0UA, UK.
Tel: 0141 945 0055. Fax: 0141 946 6173.
US address: 45150 Business Court, Suite 550, Sterling, VA 22170, USA.
Tel: 703 689 3390. Fax: 703 689 3392.

Dako Ltd, 16 Manor Courtyard, Hughenden Avenue, High Wycombe, Bucks HP13 5RE, UK.
Tel: 01494 452016. Fax: 01494 441846.
US address: 6392 Via Real, Carpinteria, CA 93013, USA.
Tel: 805 566 6655. Fax: 805 566 6688.

Digene Diagnostics Inc., 2301-B Broadbirch Drive, Silver Spring, MD 20904, USA.
Tel: 301 470 6500. Fax: 301 680 0696.

Cambridge Bioscience, 25 Signet Court, Newmarket Road, Cambridge CB5 8LA, UK.
Tel: 01223 316855. Fax: 01223 60732.

Camlab Ltd, Nuffield Road, Cambridge CB4 1TH, UK.
Tel: 01223 424222. Fax: 01223 420856.

Clontech Laboratories Inc., 4030 Fabian Way, Palo Alto, CA 94303, USA.
Tel: 415 424 8222. Fax: 415 424 1064.
(UK distributor, Cambridge BioScience.)

Costar UK Ltd, 10 The Valley Centre, Gordon Road, High Wycombe, Bucks HP13 6EQ, UK.
Tel: 01494 471207. Fax: 01494 464891.

CP Laboratories, PO Box 22, Bishop's Stortford, Herts CM23 3DX, UK.
Tel: 01279 758200. Fax: 01279 755785.

Display Systems Tandil Ltd, 31 Bedford Square, London WC1B 3SG, UK.
Tel: 0171 626 4484. Fax: 0171 626 4483.
US address: 1256 North Flores St., Suite 10, Los Angeles, CA 90069, USA.
Tel: 213 656 0894. Fax: 213 656 6332.

DNA International Inc., Suite 151, 333 South State Street, Lake Oswego, OR 97034, USA.
Tel: 800 544 4012. Fax: 503 682 6951.

Dynal Ltd., 10 Thursby Road, Croft Business Park, Bromborough, Wirral, Merseyside L62 3PW, UK.
Tel: 0151 346 1234. Fax: 0151 346 1223.
US address: 5 Delaware Drive, Lake Success, NY 11042, USA.
Tel: 800 638 9416. Fax: 516 326 3298.

Eppendorf–Netheler–Hinz GmbH, Biotech Products, Barkausenweg 1, D-22339, Hamburg, Germany.
Tel: 40 53801 0. Fax: 40 53801 593.
(UK distributor, Merck Ltd.)

Manufacturers and Suppliers

Ericomp Inc., 6044 Cornerstone Court West, Suite E, San Diego, CA 92121, USA.
Tel: 619 457 1888. Fax: 619 457 2937.

5 Prime to 3 Prime Inc., 5603 Arapahoe Road, Boulder, CO 80303, USA.
Tel: 800 533 5703. Fax: 303 440 0835.
(UK distributor, CP Laboratories.)

Flowgen Instruments Ltd, Broad Oak Enterprise Village, Broad Oak Road, Sittingbourne, Kent ME9 8AQ, UK.
Tel: 01795 429737. Fax: 01795 471185.

FMC BioProducts, 191 Thomaston Street, Rockland, ME 04841, USA.
Tel: 800 521 0390. Fax: 207 594 3491.
(UK distributor, Flowgen Instruments Ltd.)

Genetic Research Instrumentation Ltd, Gene House, Dunmow Road, Felsted, Dunmow, Essex CM6 3LD, UK.
Tel: 01371 821082. Fax: 01371 820131.

Gilson Inc., Box 27, 3000 West Beltline Highway, Middleton, WI 53562-0027, USA.
Tel: 608 836 1551. Fax: 608 831 4451.
(UK distributor, Anachem Ltd.)

GL Applied Research Inc., 142 Hawley Street, PO Box 187, Grayslake, IL 60030, USA.
Tel: 708 223 2220. Fax: 708 223 2287.

Glen Research, 44901 Falcon Place, Sterling, VA 22170, USA.
Tel: 703 437 6191. Fax: 703 435 9774.
(UK distributor, Cambio Ltd.)

Grant Instruments (Cambridge) Ltd., Barrington, Cambridge CB2 5QZ, UK.
Tel: 01763 260811. Fax: 01763 262410.
(USA distributor, Science Electronics.)

GenHunter Corp., 50 Boylston Street, Brookline, MA 32146 USA.
Tel: 617 739 6771. Fax: 617 734 5482.
(UK distributor, Bio/Gene Ltd.)

Genosys Biotechnologies, 162A Cambridge Science Park, Milton Road, Cambridge CB4 4GH, UK.
Tel: 01223 425622. Fax: 01223 425966.
US address: 1442 Lake Front Circle, Suite 185, The Woodlands, TX 77380-3600, USA.
Tel: 713 363 3693. Fax: 713 362 2212.

Gentra Systems Inc., 3905 Annapolis Lane, Minneapolis, MN 55447, USA.
Tel: 612 557 9959. Fax: 612 557 6541.
(UK distributor, Flowgen Instruments Ltd.)

GIBCO-BRL see Life Technologies.

Hoefer Scientific Instruments, PO Box 351, Newcastle-under-Lyme, Staffs ST5 0TT, UK.
Tel: 01782 617317. Fax: 01782 617346.
US address: 654 Minnesota Street, Box 77387, San Francisco, CA 94107, USA.
Tel: 800 227 4750. Fax: 415 821 1081.
(*Note added in proof*: Hoefer Scientific Instruments has now been taken over by Pharmacia Biotechnology.)

Hybaid Ltd, 111–113 Waldegrave Road, Teddington, Middx TW11 8LL, UK.
Tel: 0181 614 1000. Fax: 0181 977 0170.
(USA distributor, National Labnet Co.)

Idaho Technology Inc., 149 Chestnut Street, PO Box 50819, Idaho Falls, ID 83402, USA.
Tel: 208 524 6354. Fax: 208 524 1605.
(UK distributor, Bio/Gene Ltd.)

Manufacturers and Suppliers

Incstar Ltd, Toutley Road, Wokingham, Berks RG41 1QN, UK.
Tel: 01734 772693. Fax: 01734 792061.
US address: 1990 Industrial Bvde, Stillwater, MN 55082, USA.
Tel: 612 439 9710. Fax: 612 779 7847

Integrated DNA Technologies Inc., 1710 Commercial Park, Coralville, IO 52241, USA.
Tel: 800 328 2661. Fax: 319 645 2921.

Integrated Separation Systems, 21 Strathmore Road, Natick, MA 01760, USA.
Tel: 508 655 1500. Fax: 508 655 8501.
(UK distributor, ISS-AutoGen Ltd.)

Invitrogen Corporation, 3985-B Sorrento Valley Boulevard, San Diego, CA 92121, USA.
Tel: 619 597 6200. Fax: 619 597 6201.
(UK distributor, **R&D** Systems Europe.)

Link Technologies Ltd, 2 Napier Court, Wardpark North, Cumbernauld, Glasgow G68 0LG, UK.
Tel/Fax: 01236 451299.

Merck Ltd, Merck House, Poole, Dorset BH15 1TD, UK.
Tel: 01202 669700. Fax: 01202 666536.

Microprobe Corp., 7390 Lincoln Way, Garden Grove, CA 92641, USA.
Tel: 714 894 7184. Fax: 714 891 1229.
(UK distributor, Techgen International Ltd.)

M.J. Research Inc., 24 Bridge Street, Watertown, MA 02172, USA.
Tel: 800 729 2165. Fax: 617 924 2148.
(UK distributor; Genetic Research Instrumentation Ltd.)

Molecular Bio-Products Inc, 10171 Pacific Mesa Boulevard, Suite 307, San Diego, CA 92121, USA.
Tel: 619 453 7551. Fax: 619 453 4367.
(UK distributor, Promega Ltd.)

ISS-AutoGen Ltd, Butts Farm, Potterne, Devizes, Wilts SN10 5LR, UK.
Tel: 01380 722635. Fax: 01380 722 364.

King's College School of Medicine and Dentistry, Molecular Medicine Unit, Rayne Institute, 123 Cold Harbour Lane, London SE5 9NU, UK.
Tel: 0171 326 3126. Fax: 0171 733 3877.

Kodak Clinical Diagnostics Ltd, Mandeville House, 62 The Broadway, Amersham, Bucks HP7 0HJ, UK.
Tel: 01494 431717. Fax: 01494 725301.

Life Sciences International (UK) Ltd, Unit 5, The Ringway Centre, Edison Road, Basingstoke, Hants RG21 2YH, UK.
Tel: 01256 817282. Fax: 01256 817292.

Life Technologies Ltd, European Division, PO Box 35, 3 Fountain Drive, Inchinnan Business Park, Paisley PA5 9RF, UK.
Tel: 0141 814 6100. Fax: 0141 887 1167.
US address: PO Box 6009, Gaithersburg, MD 20877, USA.
Tel: 301 840 8000. Fax: 301 670 8539.

National Labnet Co., PO Box 841, Woodbridge, NJ 07095, USA.
Tel: 908 549 2100. Fax: 908 549 2120.

NBL Gene Sciences Ltd, South Nelson Industrial Estate, Cramlington, Northumb. NE23 9HL, UK.
Tel: 01670 733015. Fax: 01670 730454.

New England Biolabs, 67 Knowl Piece, Wilbury Way, Hitchin, Herts SG4 0TY, UK.
Tel: 01462 420616. Fax: 01462 421057.
US address: 32 Tozer Road, Beverley, MA 01915-5599, USA.
Tel: 508 927 5054. Fax: 508 921 1350.

Novagen Inc., 597 Science Drive, Madison, WI 53711, USA.
Tel: 608 238 6110. Fax: 608 238 1388.
(UK distributors, NBL Gene Sciences Ltd and AMS Biotechnology Ltd.)

Manufacturers and Suppliers

Oncogene Science Inc., 106 Charles Lindbergh Boulevard, Uniondale, NY 11553-3649, USA.
Tel: 516 222 0023. Fax: 516 222 0114.
(UK distributor, Cambridge Bioscience.)

Operon Technologies Inc., 1000 Atlantic Ave., Suite 108, Alameda, CA 94501, USA.
Tel: 510 865 8644. Fax: 510 865 5255.
(UK distributor, V.H. Bio Ltd.)

OSWEL, Medical and Biological Science Building, University of Southampton, Bolderwood, Bassett Crescent East, Southampton SO16 7TX, UK.
Tel: 01703 592974. Fax: 0800 136145; 01703 592991.

PanVera Corp., 565 Science Drive, Madison, WI 53711, USA.
Tel: 608 233 9450. Fax: 608 233 3007.

Perkin-Elmer, see Applied Biosystems.

Promega Ltd, Delta House, Enterprise Road, Chilworth Research Centre, Southampton SO1 7NS, UK.
Tel: 0800 760225. Fax: 01703 767014.
US address: 2800 Woods Hollow Road, Madison, WI 53711-5399, USA.
Tel: 800 356 9526. Fax: 608 273 6967.

Qiagen Inc., Chatsworth, CA 91311, USA.
Tel: 818 718 9870. Fax: 818 718 2956.
(UK distributor, Hybaid Ltd.)

R&D Systems Europe Ltd, 4–10 The Quadrant, Barton Lane, Abingdon, Oxon OX14 3YS, UK.
Tel: 01235 531074. Fax: 01235 533420.

Research Genetics, 2130 Memorial Parkway SW, Huntsville, AL 35801, USA.
Tel: 800 533 4363. Fax: 205 536 9016.
UK Tel: 01800 891393.

PGC Scientific, PO Box 15, Bristol BS99 5NN, UK.
Tel: 01800 960515. Fax: 01800 960516.

Pharmacia Biotechnology, 23 Grosvenor Road, St Albans, Herts, UK.
Tel: 01727 814000. Fax: 01727 814020.
US address: 800 Centennial Avenue, PO Box 1327, Piscataway, NJ 08855-1327, USA.
Tel: 201 457 8000. Fax: 201 457 0557.

Pharmingen, 11555 Sorrento Valley Road, San Diego, CA 92121, USA.
Tel: 619 792 5730. Fax: 619 792 5238.
(UK distributor, Cambridge Bioscience.)

Prime Synthesis Inc., 2 New Road, Suite 126, Aston, PA 19014, USA.
Tel: 215 558 5920. Fax: 215 558 5923.
(UK distributor, Advanced Biotechnologies Ltd.)

Robbins Scientific, Suite B, Greville Court, 1665 High Street, Knowle, Solihull, W. Mids B93 0LL, UK.
Tel: 01564 775 525. Fax: 01564 775 759.
US address: 814 San Aleso Avenue, Sunnyvale, CA 94086, USA.
Tel: 408 734 8500. Fax: 408 734 0300.

Roche Diagnostic Systems, PO Box 8, Welwyn Garden City, Herts AL7 3AY, UK.
Tel: 01707 366000. Fax: 01707 373556.
US address: 1080 US Highway 202, Branchburg, NJ 08876-1760, USA.
Tel: 908 253 7200. Fax: 908 253 7652/7646.

Sarstedt Ltd, 68 Boston Road, Beaumont Leys, Leicester LE4 1AW, UK.
Tel: 0116 2359023. Fax: 0116 2366099.
US address: Route 2, St James's Church Road, PO Box 468, Newton, NC 28658, USA.
Tel: 704 465 4000. Fax: 704 465 4003.

Manufacturers and Suppliers

Science Electronics, PO Box 986, Dayton, OH 45401-0986, USA.
Tel: 513 859 5555. Fax: 513 859 7930.

Scotlab Ltd, Kirkshaws Road, Coatbridge, Lanarks ML5 8AD, UK.
Tel: 01236 449330. Fax: 01236 449329.

Severn Biotech Ltd, Unit 2, Park Lane, Stourport Road, Kidderminster, Worcs. DY11 6TJ, UK.
Tel: 01562 825286. Fax: 01562 825284.

Sigma Chemical Company, Fancy Road, Poole, Dorset BH17 7NH, UK.
Tel: 0800 373731. Fax: 0800 378785.
US address: PO Box 14508, St Louis, MO 63178, USA.
Tel: 800 325 3010. Fax: 800 325 5052.

Sorenson Bioscience Inc., 6507 South 400 West, Salt Lake City, UT 84107 USA.
Tel: 801 266 9334. Fax: 801 262 0433.
(UK distributors, Costar UK Ltd and Bioquote Ltd.)

Stratech Scientific Ltd, 61–63 Dudley Street, Luton, Beds LU2 0NP, UK.
Tel: 01582 481884. Fax: 01582 481895.

Stuart Scientific Co. Ltd, Holmethorpe Avenue, Redhill, Surrey RH1 2NB, UK.
Tel: 01737 766431. Fax: 01737 765952.

Takara Shuzo Co. Ltd (USA distributor, PanVera Corp.; UK distributor, R&D Systems Europe Ltd.)

Techgen International Ltd, Suite 8, 50 Sullivan Road, London SW6 3DX, UK.
Tel: 0171 371 5922. Fax: 0171 371 0496.

Techne Ltd, Duxford, Cambridge CB2 4PZ, UK.
Tel: 01223 832401. Fax: 01223 836838.
US address: 3700 Brunswick Pike, Princeton, NJ 08540, USA.
Tel: 609 452 9275. Fax: 609 987 8177.

Sorin Biomedica SpA, via Crescentino, 13040 Saluggia VC, Italy.
Tel: 161 487373. Fax: 161 487642.
(USA distributor, Incstar Corp., UK distributor, Incstar Ltd.)

Stratagene Ltd, 140 Cambridge Science Park, Milton Road, Cambridge CB4 4GF, UK.
Tel: 01223 420955. Fax: 01223 420234.
US address: 11011 North Torrey Pines Road, La Jolla, CA 92037, USA.
Tel: 800 424 5444. Fax: 619 535 5430.

V.H. Bio Ltd, PO Box 7, Gosforth, Newcastle upon Tyne NE3 4DB, UK.
Tel: 0191 492 0022. Fax: 0191 410 0916.

REFERENCES

Preface

1. Newton, C.R. and Graham, A. (1994) *PCR*. BIOS Scientific Publishers, Oxford.

Chapter 1

1. Saiki, R.K., Scharf, S., Faloona, F., Mullis, K.B., Horn, G.T., Erlich, H.A. and Arnheim, N. (1985) *Science* **230**, 1350.
2. Mullis, K.B., Faloona, F., Scharf, S., Saiki, R., Horn, G. and Erlich, H.A. (1986) *Cold Spring Harbor Symp. Quant. Biol.* **51**, 263.
3. Scharf, S.J., Horn, G.T. and Erlich, H.A. (1986) *Science* **233**, 1076.
4. Saiki, R.K., Gelfand, D.H, Stoffel, S., Scharf, S.J., Higuchi, R., Horn, G.T., Mullis, K.B. and Erlich, H.A. (1988) *Science* **239**, 487.
   (1993) *Trends Genet.* **9**, 403.
21. Arnot, D.E., Roper, C. and Sultan, A.A. (1994) *Parasitol. Today* **10**, 324.
22. Crane, J.K. and Ericsson, C.D. (1991) *Curr. Opin. Infect. Dis.* **4**, 84.
23. Lyons, J. (1992) *Cancer* **69**, 1527.
24. Rettinger, S.D., Hafenrichter, D.G. and Flye, M.W. (1993) *Arch. Surg.* **128**. 1253.
25. Crisan, D. and Mattson, J.C. (1993) *DNA Cell Biol.* **12**, 455.
26. Smith, K.L. and Dunstan, R.A. (1993) *Br. J. Haematol.* **84**, 187.
27. Garson, J.A. (1994) *FEMS Microbiol. Rev.* **14**, 229.
28. Newton, C.R. (1995) in McPherson, M.J., Hames, B.D. and Taylor, G.R. (eds) *PCR-II: a Practical Approach*. Oxford University Press, Oxford.
29. The, T.H., van der Ploeg, M., Vlieger, A.M., van der Giessen, M. and van Son, W.J. (1992) *Transplantation* **54**, 193.
30. Terach, T. (1993) *J. Plant Res.* **106**, 75.

5. Wright, P.A. and Wynford-Thomas, D. (1990) *J. Pathol.* **162,** 99.
6. Bej, A.K., Mahbubani, M.H. and Atlas, R.M. (1991) *Crit. Rev. Biochem. Mol. Biol.* **26,** 301.
7. Erlich, H.A., Gelfand, D. and Sninsky, J.J. (1991) *Science* **252,** 1643.
8. Arnheim, N. and Erlich, H. (1992) *Ann. Rev. Biochem.* **61,** 131.
9. O'Garra, A. and Vieira, P. (1992) *Curr. Opin. Immunol.* **4,** 211.
10. So, A.G. and Downey, K.M. (1992) *Crit. Rev. Biochem. Mol. Biol.* **27,** 129.
11. Ellingboe, J. and Gyllensten, U.B. (eds) (1992) *The PCR Technique: DNA Sequencing.* Eaton Publishing, Natick, MA.
12. Bevan, I.S., Rapley, R. and Walker, M.R. (1992) *PCR Meth. Appl.* **1,** 222.
13. Pääbo, S. (1991) *PCR Meth. Appl.* **1,** 107.
14. Erlich, H.A. and Arnheim, N. (1992) *Ann. Rev. Genet.* **26,** 479.
15. Kocher, T.D. (1992) *PCR Meth. Appl.* **1,** 217.
16. Blake, E., Crim, D., Mihalovich, J., Higuchi, R., Walsh, P.S. and Erlich, H. (1992) *J. Forensic Sci.* **37,** 700.
17. Boehm, C.D. (1989) *Clin. Chem.* **35,** 1843.
18. Vosberg, H.P. (1989) *Hum. Genet.* **83,** 1.
19. Forrest, S. and Cotton, R.G.H. (1990) *Mol. Biol. Med.* **7,** 451.
20. Dianzani, I., Camaschella, C., Ponzone, A. and Cotton, R.G.H.

31. Koch, W.H., Payne, W.L., Wentz, B.A. and Cebula, T.A. (1993) *Appl. Env. Microbiol.* **59,** 556.
32. Furrer, B., Candrian, U., Hofelein, Ch., and Luthy, J. (1991) *J. Appl. Bacteriol.* **70,** 372.
33. Candrian, U., Furrer, B., Hofelein, Ch., Meyer, R., Jermini M. and Luthy, J. (1991) *Int. J. Food Microbiol.* **12,** 339.
34. Palmer, C.J., Tsai, Y., Paszko-Kolva, C., Mayer, C. and Sangermano, L.R. (1993) *Appl. Env. Microbiol.* **59,** 3618.
35. Steffan, R.J. and Atlas, R.M. (1991) *Ann. Rev. Microbiol.* **45,** 137.
36. Bej, A.K. and Mahbubani, M.H. (1992) *PCR Meth. Appl.* **1,** 151.
37. D' Aquila, R.T., Bechtel, L.J., Videler, J.A., Eron, J.J., Gocczyca, P. and Kaplan, J.C. (1991) *Nucl. Acids Res.* **19,** 3749.
38. Chou, Q., Russell, M., Birch, D.E., Raymond, J. and Bloch, W. (1992) *Nucl. Acids Res.* **20,** 1717.
39. Kai, M., Kamiya, S., Sawamura, S., Yamamoto, T. and Ozawa, A. (1991) *Nucl. Acids Res.* **19,** 4652.
40. Nakagomi, O., Oyamada, H. and Nakagomi, T. (1991) *Mol. Cell. Probes* **5,** 285
41. Kwok, S. and Higuchi, R. (1989) *Nature* **339,** 237.
42. Longo, M.C., Berninger, M.S. and Hartley, J.L. (1990) *Gene* **93,** 125.

References

43. Pang, J., Modlin, J. and Yolken, R. (1992) *Mol. Cell. Probes* **6**, 251.
44. Ou, C.-Y., Moore, J.L. and Schochetman, G. (1991) *BioTechniques* **10**, 442.
45. Pao, C.C., Hor, J.J., Tsai, P.L. and Horng, M.Y. (1993) *Mol. Cell. Probes* **7**, 217.
46. Cimino, G.D., Metchette, K.C., Tessman, J.W., Hearst, J.E. and Isaacs, S.T. (1991) *Nucl. Acids Res.* **19**, 99.
47. Sheng Zhu, Y., Isaacs, S.T., Cimino, G.D. and Hearst, J.E. (1991) *Nucl. Acids Res.* **19**, 2511.
48. Deragon, J.-M., Sinnett, D., Mitchell, G., Potier, M. and Labuda, D. (1990) *Nucl. Acids Res.* **18**, 6149.
49. DeFilippes, F.M. (1991) *BioTechniques* **10**, 26.
50. Dougherty, R.M., Phillips, P.E., Gibson S. and Young, L. (1993) *J. Virol. Meth.* **41**, 235.
51. Aslanzadeh, J. (1993) *Mol. Cell. Probes* **7**, 145.
52. Meier, A., Persing, D.H., Finken, M. and Bottger, E. (1993) *J. Clin. Microbiol.* **31**, 646.
53. Rys, P.N. and Persing, D.H. (1993) *J. Clin. Microbiol.* **31**, 2356.
54. Furrer, B., Candrian, U., Wieland, P. and Lüthy, J. (1990) *Nature* **346**, 324.
55. Zhu, Y.S., Isaacs, S.T., Cimino, G.D. and Hearst, J.E. (1991) *Nucl. Acids Res.* **19**, 2511.
56. Nicholls, R.D., Chakravati, A. and Ledbetter, D.H. (1993) *Hum. Mol. Genet.* 2, 143.
67. Vandenvelde, C. and Vanbeers, D. (1993) *J. Med. Virol.* **39**, 273.
68. Way, J.S., Josephson, K.L., Pillai, S.D., Abbaaszadegan, M., Gerba, C.P. and Pepper, I.L. (1993) *Appl. Env. Microbiol.* **59**, 1473.
69. O'Keefe, D.S. and Dobrovic, A. (1993) *Hum. Mutation* **2**, 67.
70. Munne, S., Tang, Y.X., Grifo, J., Rosenwaks, Z. and Cohen, J. (1994) *Fert. Ster.* **61**, 111.
71. Edwards, M.C. and Gibbs, R.A. (1994) *PCR Meth. Appl.* **3**, S65.
72. Plikaytis, B.B., Marden, J.L., Crawford, J.T., Woodley, C.L., Butler, W.R. and Shinnick, T.M. (1994) *J. Clin. Microbiol.* **32**, 1542.
73. Adeyefa, C.A.O., Quayle, K. and McCauley, J.W. (1994) *Virus Res.* **32**, 391.
74. Geha, D.J., Uhl, J.R., Gustaferro, C.A. and Persing, D.H. (1994) *J. Clin. Microbiol.* **32**, 1768.
75. Kainz, P., Schmiedlechner, A. and Strack, H.B. (1992) *Anal. Biochem.* **202**, 46.
76. Barnes, W.M. (1994) *Proc. Natl Acad. Sci. USA* **91**, 2216.
77. Foord, O.S. and Rose, E.A. (1994) *PCR Meth. Appl.* **3**, S149.
78. Cheng, S., Fockler, C., Barnes, W.M. and Higuchi, R. (1994)

56. Chamberlain, J.S., Gibbs, R.A., Ranier, J.E., Nguyen, P.N. and Caskey, C.T. (1988) *Nucl. Acids Res.* **16**, 11141.
57. Beggs, A.H., Koenig, M., Boyce, F.M. and Kunkel, L. (1990) *Hum. Genet.* **86**, 45.
58. Bej, A.K., Mahbubani, M.H., Miller, R., Dicesare, J.L., Hahh, L. and Atlas, R.M. (1990) *Mol. Cell. Probes* **4**, 353.
59. Richards, R.I., Holman, K., Lane, S., Sutherland, G.R. and Callen, D.F. (1991) *Genomics* **10**, 1047.
60. Runnebaum, I.B., Nagarajan, M., Bowman, M., Soto, D. and Sukumar, S. (1991) *Proc. Natl Acad. Sci. USA* **88**, 10657.
61. Soler, C., Allibert, P., Chardonnet, Y., Cros, P., Mandrand, B. and Thivolet, J. (1991) *J. Virol. Meth.* **35**, 143.
62. Ferrie, R.M., Schwarz, M.J., Robertson, N.H., Vaudin, S., Super, M., Malone, G. and Little, S. (1992) *Am. J. Hum. Genet.* **51**, 251.
63. Lohmann, D., Horsthemke, B., Gillessenkaesbach, G., Stefani, F.H. and Hofler, H. (1992) *Hum. Genet.* **89**, 49.
64. Burgart, L.J., Robinson, R.A., Heller, M.J., Iakoubova, O.K. and Cheville, J.C. (1992) *Mod. Path.* **5**, 320.
65. Levinson, G., Fields, R.A., Harton, G.L., Palmer, F.T., Maddalena, A., Fugger, E.F. and Schulman, J.D. (1992) *Hum. Reprod.* **7**, 1304.
66. Mutirangura, A., Greenberg, F., Butler, M.G., Malcolm, S., *Proc. Natl Acad. Sci. USA* **91**, 5695.
79. Cheng, S., Chang, S.-Y., Gravitt, P. and Respess, R. (1994) *Nature* **369**, 684.
80. Hengen, P.N. (1994) *Trends Biol. Sci.* **19**, 341.
81. Barnes, W.M. (1994) *Trends Biol. Sci.* **19**, 342.
82. Scharf, S.J., Horn, G.T. and Erlich, H.A. (1986) *Science* **233**, 1076.
83. Bhat, G.J., Lodes, M.J., Myler, P.J. and Stuart, K.D. (1991) *Nucl. Acids Res.* **19**, 398.
84. Liu, Z. and Schwartz, L.M. (1992) *BioTechniques* **12**, 29.
85. Boyd, A. C. (1993) *Nucl. Acids Res.* **21**, 817.
86. Higuchi, R., Krummel, B., Saiki, R.K. (1988) *Nucl. Acids Res.* **16**, 7351.
87. Horton, R.M., Hunt, H.D., Ho, S.N., Pullen, J.K. and Pease, L.R. (1989) *Gene* **77**, 61.
88. Yon, J. and Fried, M. (1989) *Nucl. Acids Res.* **17**, 4895.
89. Cao, Y. (1990) *Technique* **2**, 109.
90. Daugherty, B.L., DeMartino, J.A., Law, M.-F., Kawka, D.W., Singer, I.I. and Mark, G.E. (1991) *Nucl. Acids Res.* **19**, 2471.
91. Ho, S.F., Pullen, J.K., Horton, R.M., Hunt, H.D. and Pease, L.R. (1989) *DNA Prot. Eng. Tech.* **2**, 50.
92. Marchuk, D., Drumm, M., Saulino, A. and Collins, F.S. (1991) *Nucl. Acids Res.* **19**, 1154.

References

93. Kovalic, D., Kwak, J.H. and Weisblum, B. (1991) *Nucl. Acids Res.* **19,** 4560.
94. Zimmer, W. (1993) *Nucl. Acids Res.* **21,** 773.
95. Kaufman, D.L. and Evans, G.A. (1990) *BioTechniques* **9,** 304.
96. Aslanidis, C. and deJong, P.J. (1990) *Nucl. Acids Res.* **18,** 6069.
97. Haun, R.S., Serventi, I.M. and Moss, J. (1992) *BioTechniques* **13,** 515.
98. Haun. R.S. and Moss, J. (1992) *Gene* **112,** 37.
99. Nisson P.E., Rashtchian, A. and Watkins, P.C. (1991) *PCR Meth. Appl.* **1,** 120.
100. Rashtchian, A., Buchman, G.W., Schuster, D.M. and Berninger, M. (1991) *Anal. Biochem.* **206,** 91.
101. Oliner, J.D., Kinzler, K.W. and Vogelstein, B. (1993) *Nucl. Acids Res.* **21,** 5192.
102. Hsiao, K. (1993) *Nucl. Acids Res.* **21,** 5528.
103. Trout, A.B., McHeyzer-Williams, M.G., Pulendran, B. and Nossal, G.J.V. (1992) *Proc. Natl Acad. Sci. USA* **89,** 9823.
104. Frohman, M.A., Dush, M.K. and Martin, G.R. (1988) *Proc. Natl Acad. Sci. USA* **85,** 8998.
105. Belyavsky, A., Vinogradova, T. and Rajewsky, K. (1989) *Nucl. Acids Res.* **17,** 2919.
106. Frohman, M.A. and Martin, G.R. (1989) *Technique* **1,** 165.
118. Owczarek, C.M., Enriquez-Harris, P. and Proudfoot, N.J. (1992) *Nucl. Acids Res.* **20,** 851.
119. Tan, S.S. and Weis, J.H. (1992) *PCR Meth. Appl.* **2,** 137.
120. Buchman, G.W., Schuster, D.M. and Rashtchian, A. (1993) *PCR Meth. Appl.* **3,** 28.
121. Liang, P. and Pardee, A.B. (1992) *Science* **257,** 967.
122. Bauer, D., Müller, H., Reich, J., Riedel, H., Ahrenkiel, V., Warthoe, P. and Strauss, M. (1993) *Nucl. Acids Res.* **21,** 4272.
123. Heniford, B.W., Shum-Siu, A., Leonberger, M. and Hendler, F. J. (1993) *Nucl. Acids Res.* **21,** 3159.
124. Jayaraman, K., Shah, J. and Fyles, J. (1989) *Nucl. Acids Res.* **17,** 4403.
125. Barnett, R.W. and Erfle, H. (1990) *Nucl. Acids Res.* **18,** 3094.
126. Jayaraman, K., Fingar, S.A., Shah, J. and Fyles, J. (1991) *Proc. Natl Acad. Sci. USA* **88,** 4084.
127. Ciccarelli, R.B, Gunyuzlu, P., Huang, J., Scott, C. and Oakes, F. T. (1991) *Nucl. Acids Res.* **19,** 6007.
128. Sandhu, G.S., Aleff, R.A. and Kline, B.C. (1992) *BioTechniques* **12,** 14.
129. Michaels, M.L., Hsiao, H.M. and Miller, J.H. (1992) *BioTechniques* **12,** 45.
130. Prodromou, C. and Pearl, L.H. (1992) *Prot. Eng.* **5,** 827.
131. Majumder, K. (1992) *Gene* **110,** 89.

References

107. Ohara, O., Dorit, R.L. and Gilbert W. (1989) *Proc. Natl Acad. Sci. USA* **86**, 5673.
108. Loh, E.Y., Elliott, J.F., Cwirla, S., Lanier, L.L. and Davis, M.M. (1989) *Science* **243**, 217.
109. Jain, R., Gomer, R.H. and Murtagh, J.J. (1992) *BioTechniques* **12**, 58.
110. Kawasaki, E.S., Clark, S.S., Coyne, M.Y., Smith, S.D., Champlin, R., White, O.N. and McCormick, F.P. (1988) *Proc. Natl Acad. Sci. USA* **85**, 5698.
111. Wang, A.M., Doyle, M.V. and Mark, D.F. (1989) *Proc. Natl Acad. Sci. USA* **86**, 9717.
112. Becker-André, M. and Hahlbrock, K. (1989) *Nucl. Acids Res.* **17**, 9437.
113. Chelly, J., Montarras, D., Pinset, C., Berwald-Netter, Y. and Kaplan, J.C. (1990) *Eur. J. Biochem.* **187**, 691.
114. Gilliland, G., Perrin, S., Blanchard, K. and Bunn, H.F. (1990) *Proc. Natl Acad. Sci. USA* **87**, 2725.
115. Sam-Singer, J., Robison, M.O., Bellvue, A.R., Simon, M.I. and Riggs, A.D. (1990) *Nucl. Acids Res.* **18**, 1255.
116. Murphy, L.D., Herzog, C.E., Ruddick, J.B, Fojop, A.Y. and Bates, S.E. (1990) *Biochem.* **29**, 10351.
117. Robinson, J.M. and Simon, M.I. (1991) *Nucl. Acids Res.* **19**, 1557.
132. Jayaraman, K. and Puccini, C.J. (1992) *BioTechniques* **12**, 392.
133. Ye, Q.Z., Johnson, L.L. and Baragi, V. (1992) *Biochem. Biophys. Res. Commun.* **186**, 143.
134. Graham, R.W., Atkinson, T., Kilburn, D.G., Miller, R.C. and Warren, R.A. (1993) *Nucl. Acids Res.* **21**, 4923.
135. Higuchi, R., Krummel, B. and Saiki, R.K. (1988) *Nucl. Acids Res.* **16**, 7351.
136. Kunkel, T.A., Roberts, J.D. and Zakour, R.A. (1988) *Methods Enzymol.* **154**, 367.
137. Vallette, F., Mege, E., Reiss, A. and Adesnik, M. (1989) *Nucl. Acids Res.* **17**, 723.
138. Ho, S.N., Hunt, H.D., Horton, R.M., Pullen, J.K. and Pease, L.R. (1989) *Gene* **77**, 51.
139. Clackson, T.P. and Winter, G. (1989) *Nucl. Acids Res.* **17**, 10163.
140. Kahn, S.M., Jiang, W., Borner, C., O'Driscoll, K. and Weinstein, I.B. (1990) *Technique* **2**, 27.
141. Landt, O., Grunert, H.-P. and Hahn, U. (1990) *Gene* **96**, 125.
142. Perrin, S. and Gilliland, G. (1990) *Nucl. Acids Res.* **18**, 7433.
143. Shyamala, V. and Ames, G.F.-L. (1991) *Gene* **97**, 1.
144. Kuipers, O.P., Boot, H.J. and deVos, W.M. (1991) *Nucl. Acids Res.* **19**, 4558.
145. Ito, W., Ishiguro, H. and Kurosawa, Y. (1991) *Gene* **102**, 67.

146. Diaz, J.J., Rhoads, D.D. and Roufa, D.J. (1991) *BioTechniques* **11**, 204.
147. Hall, L. and Emery, D.C. (1991) *Prot. Eng.* **4**, 601.
148. Imai, Y., Matsushima, Y., Sugimara, T. and Terada, M. (1991) *Nucl. Acids Res.* **19**, 2785.
149. Sharrocks, A.D. and Shaw, P.E. (1992) *Nucl. Acids Res.* **20**, 1147.
150. Spee, J.H., de Vos, W.M. and Kuipers, O.P. (1993) *Nucl. Acids Res.* **21**, 777.
151. Cadwell, R.C. and Joyce, G.F. (1992) *PCR Meth. Appl.* **2**, 28.
152. Merino, E., Osuna, J., Bolivar, F. and Soberon, X. (1992) *BioTechniques* **12**, 508.
153. Ogel, Z.B. and McPherson, M.J. (1992) *Prot. Eng.* **5**, 467.
154. Stemmer, W.P. and Morris, S.K. (1992) *BioTechniques* **13**, 214.
155. Rashchian, A., Thornton, C.G. and Heidecker, G. (1992) *PCR Meth. Appl.* **2**, 124.
156. Deng, W.P. and Nickoloff, J.A. (1992) *Anal. Biochem.* **200**, 81.
157. Ray, F.A. and Nickoloff, J.A. (1992) *BioTechniques* **13**, 342.
158. Stappert, J., Wirsching, J. and Kemler, R. (1992) *Nucl. Acids Res.* **20**, 624.
159. Winter, G. (1993) *Curr. Opin. Immunol.* **5**, 253.
160. Skerra, A. (1993) *Curr. Opin. Immunol.* **5**, 256.
161. Smith, K.D., Valenzuela, A., Vigna, J.L., Aalbers, K. and Lutz, C.T. (1993) *PCR Meth. Appl.* **2**, 253.
175. Tropak, M.B. and Roder, J.C. (1994) *J. Neurochem.* **62**, 854.
176. Wong, C., Dowling, C.E., Saiki, R.K., Higuchi, R.G., Erlich, H.A. and Kazazian, H.H. (1987) *Nature* **330**, 384.
177. Wrischnik, L.A., Higuchi, R.G., Stoneking, M., Erlich, H.A., Arnheim, N. and Wilson, A.C. (1987) *Nucl. Acids Res.* **15**, 529.
178. McMahon, G., Davis, E. and Wogan, G.N. (1987) *Proc. Natl Acad. Sci. USA* **84**, 4974.
179. Engelke, D.R., Hoener, P.A. and Collins, F.S. (1988) *Proc. Natl Acad. Sci. USA* **85**, 544.
180. Gyllensten, U.B. and Erlich, H.A. (1988) *Proc. Natl Acad. Sci. USA* **85**, 7652.
181. Newton, C.R., Kalsheker, N., Graham, A., Powell, S., Gammack, A., Riley, J. and Markham, A.F. (1988) *Nucl. Acids Res.* **16**, 8233.
182. Innis, M.A., Myambo, K.B., Gelfand, D.H. and Brow, M.A.D. (1988) *Proc. Natl Acad. Sci. USA* **85**, 9436.
183. Nakamaye, K.L., Gish, G., Eckstein, F. and Vosberg, H.P. (1988) *Nucl. Acids Res.* **16**, 9947.
184. Stahl, S., Hultman, T., Olsson, A., Moks, T. and Uhlén, M. (1988) *Nucl. Acids Res.* **16**, 3025.
185. Mitchell, L.G. and Merril, C.R. (1989) *Anal. Biochem.* **178**, 239.
186. Hultman, T., Stahl, S., Hornes, E. and Uhlén, M. (1989) *Nucl. Acids Res.* **17**, 4937.

162. Marini, F., Naeem, A. and Lapeyre, J.N. (1993) *Nucl. Acids Res.* **21**, 2277.
163. Verhasselt, P., Reekmans, M.J. and Volckaert, G. (1993) *Genet. Anal. Tech. Appl.* **10**, 16.
164. Li, X.M. and Shapiro, L.J. (1993) *Nucl. Acids Res.* **21**, 3745.
165. Armengaud, J. and Jouanneau, Y. (1993) *Nucl. Acids Res.* **21**, 4424.
166. Tessier, D.C. and Thomas, D.Y. (1993) *BioTechniques* **15**, 498.
167. Chen, J.J., Shih, N.L., Hsu, K.H. and Liew, C.C. (1993) *Mol. Cell. Biochem.* **124**, 81.
168. Barettino, D., Feibenbutz, M., Valcarcel, R. and Stunnenberg, H.G. (1994) *Nucl. Acids Res.* **22**, 541.
169. Morris, J.A. and McIvor, R.S. (1994) *Biochem. Pharmacol.* **47**, 1207.
170. Picard, V., Ersdal Badju, E., Lu, A. and Bock, S.C. (1994) *Nucl. Acids Res.* **22**, 2587.
171. Corcoran, A.E., Barret, K., Turner, M., Brown, A., Kissonerghis, A.M., Gadnell, M., Gray, P.W., Chernajovsky, Y. and Feldmann, M. (1994) *Eur. J. Biochem.* **223**, 831.
172. Brudnak, M. and Miller, K.S. (1993) *BioTechniques* **14**, 66.
173. Browning, K.S. (1989) *Amplifications* **3**, 14.
174. Kain, K.C., Orlandi, P.A. and Lanar, D.E. (1991) *BioTechniques* **10**, 366.

187. Kusukawa, N., Uemori, T., Sada, K. and Kato, I. (1990) *BioTechniques* **9**, 66.
188. Gonzalez-Cadavid, N., Gatti, R.A. and Neuwirth, H. (1990) *Anal. Biochem.* **191**, 359.
189. Hultman, T., Bergh, S., Moks, T. and Uhlén, M. (1991) *BioTechniques* **10**, 84.
190. Kaneoka, H., Lee, D.R., Hsu, K.-C., Sharp, G.C. and Hoffman, R.W. (1991) *BioTechniques* **10**, 30.
191. Murray, V. (1989) *Nucl. Acids Res.* **17**, 8889.
192. McCabe, P.C. (1990) in *PCR Protocols: a Guide to Methods and Applications* (M.A. Innis, D.H. Gelfand, J.J. Sninsky and T.J. White, eds), p. 76. Academic Press, San Diego.
193. Lee, J.S. (1991) *DNA Cell Biol.* **10**, 67.
194. Stoflet, E.S., Koeberl, D.D., Sarkar, G. and Sommer, S.S. (1988) *Science* **239**, 491.
195. Cliby, W., Sarkar, G., Ritland, S.R., Hartmann, L., Podratz, K.C. and Jenkins, R.B. (1993) *Gynecol. Oncol.* **50**, 34.
196. Sarkar, G. and Sommer, S.S. (1989) *Science* **244**, 331.
197. Pääbo, S., and Wilson, A.C. (1988) *Nature* **334**, 387.
198. Pääbo, S., Gifford, J.A. and Wilson, A.C. (1988) *Nucl. Acids Res.* **16**, 9775.
199. Lawlor, D.A., Dickel, C.D., Hauswirth, W.W. and Parham, P. (1991) *Nature* **349**, 785.

References

200. Cano, R.J., Poinar, H.N. and Poinar, G.O., Jr (1992) *Med. Sci. Res.* **20**, 619.
201. De Salle, R., Gatesy, J., Wheeler, W. and Grimaldi, D. (1992) *Science* **257**, 1933.
202. Poinar, H.N., Cano, R.J. and Poinar, G.O., Jr. (1993) *Nature* **363**, 677.
203. Cano, R.J., Poinar, H.N., Pieniazek, N.J., Acra, A. and Poinar, G. O., Jr (1993) *Nature* **363**, 536.
204. Höss, M., Pääbo, S. and Vereshchagin, N.K. (1993) *Nature* **370**, 333.
205. Bugawan, T.L., Horn, G.T., Long, C.M., Mickelson, E., Hansen, J.A., Ferrara, G.B., Angelini, G. and Erlich, H.A. (1988) *J. Immunol.* **141**, 4024.
206. Tiercy, J.M., Gorski, J., Jeannet, M. and Mack, B. (1988) *Proc. Natl Acad. Sci. USA* **85**, 198.
207. Bugawan, T.L., Saiki, R.K., Levenson, C.H., Watson, R.W. and Erlich, H.A. (1988) *BioTechnology* **6**, 943.
208. Higuchi, R., von Beroldingen, C. H., Sensabaugh, G. F. and Erlich, H.A. (1988) *Nature* **332**, 543.
209. Fernandezvina, M., Moreas, M.E. and Stastny, P. (1991) *Hum. Immunol.* **30**, 60.
210. Lo, Y.M.D., Mehal, W.Z., Wordsworth, B.P., Chapman, R.W., Fleming, K.A., Bell, J.I. and Wainscoat, J.S. (1991) *Lancet* **338**, 65.
222. Bell, C.J. and Ecker, J.R. (1994) *Genomics* **19**, 137.
223. Zicetkiewicz, E., Rafalski, A. and Labuda, D. (1994) *Genomics* **20**, 176.
224. Adamson, R., Jones, A.S. and Field, J.K. (1994) *Oncogene* **9**, 2077.
225. King, B.L., Lichtenstein, A., Berenson, J. and Kacinski, B.M. (1994) *Am. J. Pathol.* **144**, 486.
226. Riley, J., Butler, R., Finniear, R., Jenner, D., Powell, S., Anand, R., Smith, J.C. and Markham, A.F. (1990) *Nucl. Acids Res.* **18**, 2887.
227. Arnold, C. and Hodgson, I.J. (1991) *PCR Meth. Appl.* **1**, 39.
228. Copley, C.G., Boot, C., Bundell, K. and McPheat, W.L. (1991) *BioTechnology* **9**, 74.
229. Coffey, A.J., Roberts, R.G., Green, E.D., Cole, C.G., Butler, R., Anand, R., Giannelli, F. and Bentley, D.R. (1992) *Genomics* **12**, 474.
230. Mills, K.I., Sproul, A.M., Ogilvie, D., Elvin, P., Leibowitz, D. and Burnett, A. K. (1992) *Leukemia* **6**, 481.
231. Roberts, R.G., Coffey, A.J., Bobrow, M. and Bentley, D.R. (1992) *Genomics* **13**, 942.
232. Nieuwenhuijsen, B.W., Chen, K.L., Chinault, A.C., Wang, S., Valmiki, V.H., Meershoek, E.J., van Ommen, G.J. and Fischbeck, K.H. (1992) *Hum. Mol. Genet.* **1**, 605.

211. Browning, M.J., Krausa, P., Rowan, A., Bicknell, D.C., Bodmer, J.G. and Bodmer, W.F. (1993) *Proc. Natl Acad. Sci. USA* **90**, 2842.
212. Patel, P., Lo, Y.M.D., Bell, J.I. and Wainscoat, J.S. (1993) *J. Clin. Pathol.* **46**, 1105.
213. Eliaou, J.F., Palmade, F., Avinens, O., Edouard, E., Ballaguer, P., Nicolas, J.C. and Clot, J. (1992) *Hum. Immunol.* **35**, 215.
214. Cornall, R.J., Aitman, T.J., Hearne, C.W. and Todd, J.A. (1991) *Genomics* **10**, 874.
215. Thompson, A.D., Shen, Y., Holman, K., Sutherland, G.R., Callen, D.F. and Richards, R.I. (1992) *Genomics* **13**, 402.
216. Ziegle, J.S., Su, Y., Corcoran, K.P., Nie, L., Mayrand, P.E. Hoff, L.B., McBride, L.J., Kronick, M.N. and Diehl, S.R. (1992) *Genomics* **14**, 1026.
217. Cohen, B.B., Wallace, M.R. and Crichton, D.N. (1992) *Mol. Cell. Probes* **6**, 439.
218. Lane, S.A., Taylor, G.R., Ozols, B. and Quirke, P. (1993) *J. Clin. Pathol.* **46**, 346.
219. Hauge, X.Y. and Litt, M. (1993) *Hum. Mol. Genet.* **2**, 411.
220. Gruis, N.A., Abeln, E.C., Bardoel, A.F., Devilee, P., Frants, R.R. and Cornelisse, C.J. (1993) *Br. J. Cancer* **68**, 308.
221. Morrison, K.E., Daniels, R.J., Suthers, G.K. *et al.* (1993) *Hum. Genet.* **92**, 133.
233. Gibson, R.A., Buchwald, M., Roberts, R.G. and Mathew, C.G. (1993) *Hum. Mol. Genet.* **2**, 35.
234. Kleyn, P.W., Wang, C.H., Lien, L.L. *et al.* (1993) *Proc. Natl Acad. Sci. USA* **90**, 6801.
235. Roberts, R.G., Coffey, A.J., Bobrow, M. and Bentley, D.R. (1993) *Genomics* **16**, 536.
236. Triglia, T., Peterson, M.G. and Kemp, D.J. (1988) *Nucl. Acids Res.* **16**, 8186.
237. Does, M.P., Dekker, B.M., de Groot, M.J. and Offringa, R. (1991) *Plant Mol. Biol.* **17**, 151.
238. Lo, Y.M.D., Patel, P., Newton, C.R., Markham, A.F., Flemming, K.A. and Wainscoat, J.S. (1991) *Nucl. Acids Res.*, **19**, 3561.
239. Green, I.R. and Sargan, D.R. (1991) *Gene* **109**, 203.
240. Kaluza, B., Betzl, G., Shao, H., Diamantstein, T. and Weidle, U.H. (1991) *Gene* **122**, 321.
241. Stemmer, W.P., Morris, S.K., Kautzer, C.R. and Wilson, B.S. (1993) *Gene* **123**, 1.
242. Zilberberg, N. and Gurevitz, M. (1993) *Anal. Biochem.* **209**, 203.
243. Langley, R.J., Hirsch, V.M., O'Brien, S.J., Adger-Johnson, D., Goeken, R.M. and Olmsted, R.A. (1994) *Virology* **202**, 853.
244. Ou, C.Y., Kwok, S., Mitchell, S.W., Mack, D.H., Sninsky, J.J.,

244. Krebs, J.W., Feorino, P., Warfield, D. and Schochetman, G. (1988) *Science* **239**, 295.
245. Duggan, D.B., Ehrlich, G.D., Davey, F.P., Kwok, S., Sninsky, J.J., Goldberg, J., Baltrucki, L. and Poiesz, B.J. (1988) *Blood* **71**, 1027.
246. Thiers, V., Nakahima, E., Kremsdorf, D., Mack, D., Schellekens, H., Driss, F., Goudeau, A., Wands, J., Sninsky, J. and Tiollais, P. (1988) *Lancet* **2**, 1273.
247. Shibata, D.K., Arnheim, N. and Martin, W.J. (1988) *J. Exp. Med.* **167**, 225.
248. Kellogg, D.E., Sninsky, J.J. and Kwok, S. (1990) *Anal. Biochem.* **189**, 202.
249. Aoki, S., Yarchoan, R., Thomas, R.V., Pluda, J.M., Marczyk, K., Broder, S. and Mitsuya, H. (1990) *AIDS Res. Hum. Retrovir.* **6**, 1331.
250. Piatak, M., Luk, K.-C., Williams, B. and Lifson, J.D. (1993) *BioTechniques* **14**, 70.
251. Olive, D.M. (1989) *J. Clin. Microbiol.* **27**, 261.
252. Malloy, D.C., Nauman, R.K. and Paxton, H. (1990) *J. Clin. Microbiol.* **28**, 1089.
253. Atlas, R.M. and Bej. A.K. (1990) in *PCR Protocols: a Guide to Methods and Applications* (M.A. Innis, D.H. Gelfand, J.J. Sninsky and T.J. White, eds), p. 399. Academic Press, San Diego.

266. Nuovo, G.J., Margiotta, M., MacConnell, P. and Becker, J. (1992) *Diagn. Mol. Pathol.* **1**, 98.
267. Chiu, K.P., Cohen, S.H., Morris, D.W. and Jordan, G.W. (1992) *Histochem. Cytochem.* **40**, 333.
268. O'Leary, J.J., Browne, G., Landers, R.J. *et al.* (1994) *Histochem. J.* **26**, 337.
269. Nuovo, G.J., Gallery, F., MacConnell, P., Becker, J. and Bloch, W. (1991) *Am. J. Pathol.* **139**, 1239.
270. Nuovo, G.J., Gallery, F. and MacConnell, P. (1992) *Mod. Pathol.* **5**, 444.
271. Zehbe, I., Hacker, G.W., Rylander, E., Sallstrom, J. and Wilander, E. (1992) *Anticancer Res.* **12**, 2165.
272. Kominoth, P., Long, A.A., Ray, R. and Wolfe, H.J. (1992) *Diagn. Mol. Pathol.* **1**, 85.
273. Nuovo, G.J., Gallery, F., Hom, R., MacConnell, P. and Bloch, W. (1993) *PCR Meth. Appl.* **2**, 305.
274. Gressens, P. and Martin, J.R. (1994) *J. Neuropathol. Exp. Neurol.* **53**, 127.
275. Nuovo, G.J., Becker, J., Burk, M.W., Margiotta, M., Fuhrer, J. and Steigbigel, R.T. (1994) *J. AIDS* **7**, 916.
276. Nuovo, G.J., Lidonnici, K., MacConnell, P. and Lane, B. (1993) *Am. J. Surg. Pathol.* **17**, 683.
277. Heniford, B.W., Shum, S.A., Leonberger, M. and Hendler, F.J.

254. Cousins, D.V., Wilton, S.D. and Francis, B.R. (1991) *Vet. Microbiol.* **27**, 187.
255. Wilton, S. and Cousins, D. (1992) *PCR Meth. Appl.* **1**, 269.
256. Brousseau, R., Saint-Onge, A., Préfontaine, G., Masson, L. and Cabana, J. (1993) *Appl. Env. Microbiol.* **59**, 114.
257. Jensen, M.A., Webster, J.A. and Straus, N. (1993) *Appl. Env. Microbiol.* **59**, 945.
258. Impraim, C.C., Saiki, R.K., Erlich, H.A. and Teplitz, R.L. (1987) *Biochem. Biophys. Res. Comm.* **142**, 710.
259. Shibata. D.K., Arnheim, N. and Martin, W.J. (1988) *J. Exp. Med.* **167**, 225.
260. Stanta, G. and Schneider, C. (1991) *BioTechniques* **11**, 304.
261. Greer, C.E., Lund, J.K. and Manos, M.M. (1991) *PCR Meth. Appl.* **1**, 46.
262. Jackson, D.P., Hayden, J.D. and Quirke, P. (1991) in *PCR: a Practical Approach* (M.J. McPherson, P. Quirke and G.R. Taylor, eds), p. 37. Oxford University Press, Oxford.
263. Greer, C.E., Peterson, S.J., Kiviat, N.B. and Manos, M.M. (1991) *Am. J. Clin. Pathol.* **95**, 117.
264. Nuovo, G.J., MacConnell, P., Forde, A. and Delvenne, P. (1991) *Am. J. Pathol.* **139**, 847.
265. Bagasra, O., Hauptman, S.P., Lischer, H.W., Sachs, M. and Pomerantz, R.J. (1992) *N. Engl. J. Med.* **326**, 1385.
    (1993) *Nucl. Acids Res.* **21**, 3159.
278. Chen, R.H. and Fuggle, S.V. (1993) *Am. J. Pathol.* **143**, 1527.
279. Nuovo, G.J., Gallery, F., MacConnell, P. and Braun, A. (1994) *Am. J. Pathol.* **144**, 659.
280. Kelleher, M.B., Galutira, D., Duggan, T.D. and Nuovo, G.J. (1994) *Diagn. Mol. Pathol.* **3**, 105.
281. Billadeau, D., Blackstadt, M., Greipp, P., Kyle, R.A., Oken, M.M., Kay, N. and Van Ness, B. (1991) *Blood* **78**, 3021.
282. Castiagne, S., Balitrand, N., de The, H., Dejean, A., Degos, L. and Chomienne, C. (1992) *Blood* **79**, 3110.
283. Chen, S.J., Chen, Z., Chen, A., Tong, J.H., Dong, S., Wang, Z.Y., Waxman, S. and Zelent, A. (1992) *Oncogene* **7**, 1223.
284. Kozu, T., Miyoshi, H., Shimuzu, K., Maseki, N., Asou, H., Kamada, N. and Ohki, M. (1993) *Blood* **82**, 1270.
285. Breit, T.M., Beishuizen, A., Ludwig, W.D., Mol, E.J., Adriaansen, H.J., van Wering, E.R. and van Dongen, J.J. (1993) *Leukemia* **7**, 2004.
286. Maruyama, F., Stass, S.A., Estey, E.H., Cork, A., Hirano, M., Freireich, E.J., Yang, P. and Chang, K.S. (1994) *Leukemia* **8**, 40.
287. Beishuizen, A., Verhoeven, M.A., van Wering, E.R., Hahlen, K., Hooijkaas, H. and van Dongen, J.J. (1994) *Blood* **83**, 2238.
288. Newton, C.R., Graham, A., Heptinstall, L.E., Powell, S.J.,

288. Summers, C., Kalsheker, N., Smith, J.C. and Markham, A.F. (1989) *Nucl. Acids Res.* **17**, 2503.
289. Wu, D.Y., Ugozzoli, L., Pal, B.K. and Wallace, R.B. (1989) *Proc. Natl Acad. Sci. USA* **86**, 2757.
290. Ehlen, T. and Dubeau, L. (1989) *Biochem. Biophys. Res. Commun.* **160**, 441.
291. Newton, C.R., Heptinstall, L.E., Summers, C., Super, M., Schwarz, M., Anwar, R., Graham, A., Smith, J.C. and Markham, A.F. (1989) *Lancet* **ii**, 1481.
292. Sarkar, G., Cassady, J., Bottema, D.K. and Sommer, S.S. (1989) *Anal. Biochem.* **186**, 64.
293. Newton, C.R., Schwarz, M., Summers, C., Heptinstall, L.E. Graham, A., Smith, J.C., Super, M. and Markham, A.F. (1990) *Lancet* **335**, 1217.
294. Heim, M. and Meyer, U.A. (1990) *Lancet* **336**, 529.
295. Old, J.M., Varawalla, N.Y. and Weatherall, D.J. (1990) *Lancet* **336**, 834.
296. Wagner, M., Schloesser, M. and Reiss, J. (1990) *Mol. Biol. Med.* **7**, 359.
297. Newton, C.R., Summers, C., Heptinstall, L.E, Jenner, D.E, Graham, A. and Markham, A.F. (1991) *BioTechniques* **10**, 582.
298. Wenham, P.R., Newton, C.R. and Price, W.H. (1991) *Clin. Chem.* **37**, 241.
Genet. **51**, 675.
311. Schwartz, L.S., Tarleton, J., Popovich, B., Seltzer, W.K. and Hoffman, E.P. (1992) *Am. J. Hum. Genet.* **51**, 721.
312. Kornreich, R. and Desnick, R.J. (1993) *Hum. Mutat.* **2**, 108.
313. Mansfield, E.S., Robertson, J.M., Lebo, R.V., Lucero, M.Y., Mayrand, P.E., Rappaport, E., Parrella, T., Sartore, M., Surrey, S. and Fortina, P. (1993) *Am. J. Med. Genet.* **48**, 200.
314. Cross, N.C., Melo, J.V., Feng, L. and Goldman, J.M. (1994) *Leukemia* **8**, 186.
315. Kruyer, H., Miranda, M., Volpini, V. and Estivill, X. (1994) *Prenat. Diagn.* **14**, 123.
316. Fuscoe, J.C., Nelsen, A.J. and Pilia, G. (1994) *Somat. Cell Mol. Genet.* **20**, 39.
317. Saiki, R.K., Bugawan, T.L., Horn, G.T., Mullis, K.B. and Erlich, H.A. (1986) *Nature* **234**, 163.
318. Saiki, R.K., Walsh, P.S., Levenson, C.H. and Erlich, H.A. (1989) *Proc. Natl Acad. Sci. USA* **86**, 6230.
319. Serre, J.L., Taillandier, A., Mornet, E., Simon-Bouy, B., Boué, J. and Boué, A. (1991) *Genomics* **11**, 1149.
320. Chehab, F.F. and Wall, J. (1992) *Hum. Genet.* **89**, 163.
321. Gibbs, R.A., Nguyen, P.N. and Caskey, C.T. (1989) *Nucl. Acids Res.* **17**, 2437.

299. Lo, Y.M.D., Patel, P., Mehal, W.Z., Flemming, K.A., Bell, J.I. and Wainscoat, J.S. (1992) *Nucl. Acids Res.* **20**, 1005.
300. Sommer, S.S., Groszbach, A.R. and Bottema, C.D.K. (1992) *BioTechniques* **12**, 82.
301. Ugozzoli, L. and Wallace, R.B. (1992) *Genomics* **12**, 670.
302. Bottema, C.D.K. and Sommer, S.S. (1993) *Mutat. Res.* **288**, 93.
303. Lewis, B.D., Nelson, P.V., Robertson, E.F. and Morris, C.P. (1994) *Am. J. Med. Genet.* **49**, 218.
304. Fortina, P., Conant, R., Parrella, T., Rappaport, E. Scanlin, T., Schwartz, E., Robertson, J.M. and Surrey, S. (1992) *Mol. Cell. Probes* **6**, 353.
305. Fortina, P., Conant, R., Monokian, G., Dotti, G., Parrella, T., Hitchcock, W., Kant, J., Scanlin, T., Rappaport, E., Schwartz, E. and Surrey, S. (1992) *Hum. Genet.* **90**, 375.
306. Bienvenu, T., Sebillon, P., Labie, D., Kaplan, J.C. and Beldjord, C. (1992) *Hum. Biol.* **64**, 107.
307. Fortina, P., Dotti, G., Conant, R., Monokian, G., Parrella, T., Hitchcock, W., Rappaport, E., Schwartz, E. and Surrey, S. (1992) *PCR Meth. Appl.* **2**, 163.
308. Picci, L., Anglani, F., Scarpa, M. and Zacchello, F. (1992) *Hum. Genet.* **88**, 552.
309. Abbs, S. and Bobrow, M. (1992) *J. Med. Genet.* **29**, 375.
310. Covone, A.E., Caroli, F. and Romeo, G. (1992) *Am. J. Hum.*
322. Chehab, F.F. and Kan, Y.W. (1990) *Lancet* **335**, 15.
323. Efremov, D.G., Dimovski, A.J., Janovic, L. and Efremov, G.D. (1991) *Acta Haematol.* **85**, 66.
324. Holland, P.M., Abramson, R.D., Watson, R. and Gelfand, D.H. (1991) *Proc. Natl Acad. Sci. USA* **88**, 7276.
325. Haliassos, A., Chomel, J.C., Tesson, L., Baudis, M., Kruh, J., Kaplan, J.C. and Kitzis, A. (1989) *Nucl. Acids Res.* **17**, 3606.
326. Sorscher, E.J. and Huang, Z. (1991) *Lancet* **337**, 1115.
327. Gregersen, N., Blakemore, A.I.F., Winter, V., Andresen, B., Kolvraa, S., Bolund, L., Curtis, D. and Engel, P.C. (1991) *Clin. Chim. Acta* **203**, 23.
328. Gasparini, P., Bonizzato, A., Dognini, M. and Pignatti, P.F. (1992) *Mol. Cell. Probes* **6**, 1.
329. Bal, J., Rininsland, F., Osbourne, L. and Reiss, J. (1992) *Mol. Cell. Probes* **6**, 9.
330. Stocks, J., Thorn, J.A. and Galton, D.J. (1992) *J. Lipid Res.* **33**, 853.
331. Petruzzella, V., Chen, X. and Schon, E.A. (1992) *Biochem. Biophys. Res. Comm.* **186**, 491.
332. Kogan, S.C., Doherty, M. and Gitscher, J. (1987) *N. Engl. J. Med.* **317**, 985.
333. Abbott, C.M., McMahon, J.C., Whitehouse, D.B. and Povey, S. (1988) *Lancet* **i**, 763.

334. Kuppuswamy, M.N., Hoffmann, J.W., Kasper, C.K., Spitzer, S.G., Groce, S.L. and Bajaj, S.P. (1991) *Proc. Natl Acad. Sci. USA* **88**, 1143.
335. Lin, F.H., Lin, R., Wisniewski, M., Hwang, Y.W., Grundke-Iqbal, I., Healy-Louie, G. and Iqbal, K. (1992) *Biochem. Biophys. Res. Comm.* **182**, 238.
336. Lin, F.H. and Lin, R. (1992) *Biochem. Biophys. Res. Comm.* **189**, 1202.
337. Singer, S.J. (1994) *PCR Meth. Appl.* **3**, S48.
338. Coutelle, C., Williams, C., Handyside, A., Hardy, K., Winston, R. and Williamson, R. (1989) *Br. Med. J.* **229**, 22.
339. Handyside, A.H., Kontogianni, E.H., Hardy, K. and Winston, R.M.L. (1990) *Nature* **344**, 768.
340. Navidi, W. and Arnheim, N. (1991) *Hum. Reprod.* **6**, 836.
341. Navidi, W. and Arnheim, N. (1992) *Hum. Reprod.* **7**, 288.
342. Holding, C., Bentley, D., Roberts, R., Bobrow, M. and Mathew, C. (1993) *J. Med. Genet.* **30**, 903.
343. Wu, R., Cuppens, H., Buyse, I., Decorte, R., Marynen, P., Gordts, S. and Cassiman, J. (1993) *Prenat. Diagn.* **13**, 1111.
344. Sheffield, V.C., Cox, D.R., Lerman, L.S. and Myers, R.M. (1989) *Proc. Natl Acad. Sci. USA* **86**, 232.
345. Abrams, E.S., Murdaugh, S.E. and Lerman, L.S. (1990) *Genomics* **7**, 463.
357. Nigro, V., Politano, L., Nigro, G., Romano, S.C., Molinari, A. M. and Puca, G.A. (1992) *Hum. Mol. Genet.* **1**, 517.
358. D'Amico, D., Caebone, D., Mitsudomi, T. *et al.* (1992) *Oncogene* **7**, 339.
359. Cotton, R.G.H., Rodrigues, N.R. and Campbell, R.D. (1988) *Proc. Natl Acad. Sci. USA* **85**, 4397.
360. Akli, S., Chelly, J., Lacorte, J.M., Poenaru, L. and Kahn, A. (1991) *Genomics* **11**, 124.
361. Roberts, R.G., Bobrow, M. and Bentley, D.R. (1992) *Proc. Natl Acad. Sci. USA* **89**, 2331.
362. Litjens, T., Morris, C.P., Robertson, E.F., Peters, C., vonFigura, K. and Hopwood, J.J. (1992) *Hum. Mutat.* **1**, 397.
363. Whitney, M.A., Saito, H., Jakobs, P.M., Gibson, R.A., Moses, R.E. and Grompe, M. (1993) *Nature Genetics* **4**, 202.
364. Scott, H.S., Litjens, T., Nelson, P.V., Thompson, P.R., Brooks, D.A., Hopwood, J.J. and Morris, C.P. (1993) *Am. J. Hum. Genet.* **53**, 973.
365. Williams, J.G.K., Kubelik, A.R., Livak, K.J., Rafalski, J.A. and Tingey, S.V. (1990) *Nucl. Acids Res.* **18**, 6531.
366. Welsh, J. and McClelland, M. (1991) *Nucl. Acids Res.* **18**, 7213.
367. Caetano-Anollés, G., Bassam, B.J. and Gresshoff, P.M. (1991) *BioTechnology* **9**, 553.
368. Owen, J.L. and Uyeda, C.M. (1991) *Animal Biotechnol.* **2**, 107.

346. Higuchi, M., Antonarakis, S.E., Kasch, L., Oldenberg, J., Economou-Petersen, E., Olek, K., Arai, M., Inaba, H. and Kazazian, H.H. (1991) *Proc. Natl Acad. Sci. USA* **88**, 8307.
347. Parker, S., Angelico, M.C., Laffel, L. and Krolewski, A.S. (1993) *Genomics* **16**, 245.
348. Orita, M., Suzuki, T., Sekiya, T. and Hayashi, K. (1989) *Genomics* **5**, 874.
349. Hayashi, K. (1991) *PCR Meth. Appl.* **1**, 34.
350. Spinardi, L., Mazars, R. and Theillet, C. (1991) *Nucl. Acids Res.* **19**, 4009.
351. Dockhorn-Dworniczak, B., Dworniczak, B., Brommelkamp, L., Bulles, J., Horst, J. and Bocker, W.W. (1991) *Nucl. Acids Res.* **19**, 2500.
352. Poduslo, S.E., Dean, M., Kolch, U. and O'Brien, S.J. (1991) *Am. J. Hum. Genet.* **49**, 106.
353. Mashiyama, S., Murakami, Y., Yoshimoto, T., Sekiya, T. and Hayashi, K. (1991) *Oncogene* **6**, 1313.
354. Iizuka, M., Mashiyama, S., Oshimura, M., Sekiya, T. and Hayashi, K. (1991) *Genomics* **12**, 139.
355. Sarkar, G., Yoon, H.S. and Sommer, S.S. (1992) *Nucl. Acids Res.* **20**, 871.
356. Makino, R., Yazyu, H., Kishimoto, Y., Sekiya, T. and Hayashi, K. (1992) *PCR Meth. Appl.* **2**, 10.
369. Hu, J. and Quiros, C.F. (1991) *Plant Cell Rep.* **10**, 505.
370. Fekete, A., Bantle, J.A., Halling, S.M. and Stitch, R.W. (1992) *J. Bacteriol.* **174**, 7778.
371. Pellissier-Scott, M., Haymes, K.M. and Williams, S.M. (1992) *Nucl. Acids Res.* **20**, 5493.
372. Peinado, M.A., Malkhosyan, S., Velazquez, A. and Perucho, M. (1992) *Proc. Natl Acad. Sci. USA* **89**, 1477.
373. Tibayrenc, M., Neubauer, K., Barnabé, C., Guerrini, F., Skarecky, D. and Ayala, F.J. (1993) *Proc. Natl Acad. Sci. USA* **90**, 1335.
374. Caetano-Anollés, G. (1993) *PCR Meth. Appl.* **3**, 85.
375. Ellsworth, D.L., Rittenhouse, K.D. and Honeycutt, R.L. (1993) *BioTechniques* **14**, 214.
376. Levin, I., Crittenden, L.B. and Dodgson, J.B. (1993) *Genomics* **16**, 224.
377. Lamboy, W.F. (1994) *PCR Meth. Appl.* **4**, 31.
378. Lamboy, W.F. (1994) *PCR Meth. Appl.* **4**, 38.
379. Park, Y.H. and Kohel, R.J. (1994) *BioTechniques* **16**, 652.
380. Czajka, J. and Batt, C.A. (1994) *J. Clin. Microbiol.* **32**, 1280.
381. Weber, J.L. and May, P.E. (1989) *Am. J. Hum. Genet.* **44**, 388.
382. Litt, M. and Lutty, J.A. (1989) *Am. J. Hum. Genet.* **44**, 397.
383. Spirio, L., Joslyn, G., Nelson, L., Leppert, M. and White, R. (1991) *Nucl. Acids Res.* **19**, 6348.

References

384. Richards, R.I., Holman, K., Shen, Y., Kozman, H., Harley, H., Brook, D. and Shaw, D. (1991) *Genomics* **11**, 77.
385. Morral, N. and Estivill, X. (1992) *Genomics* **13**, 1362.
386. Furlong, R.A., Goudie, D.R., Carter, N.P., Lyall, J.E., Affara, N.A. and Ferguson-Smith, N.A. (1993) *Am. J. Hum. Genet.* **52**, 1191.
387. Odelberg, S.J. and White, R. (1993) *PCR Meth. Appl.* **3**, 7.
388. Nelson, D.L., Ledbetter, S.A., Corbo, L., Victoria, M.F., Ramirez-Solis, R., Webster, T.D., Ledbetter, D.H. and Caskey, C.T. (1989) *Proc. Natl Acad. Sci. USA* **86**, 6686.
389. Silverman, G.A., Ye, R.D., Pollock, K.M., Saddler, J.E. and Korsmeyer, S.J. (1989) *Proc. Natl Acad. Sci. USA* **86**, 7485.
390. Breukel, C., Wijnen, J., Tops, C., van der Klift, H., Dauwerse, H. and Khan, P.M. (1990) *Nucl. Acids Res.* **18**, 3097.
391. Nelson, D.L., Ballabio, A., Victoria, M.F., Pieretti, M., Bies, R. D., Webster, T.D. and Caskey, C.T. (1991) *Proc. Natl Acad. Sci. USA* **88**, 6157.
392. Palazzolo, M.J., Sawyer, S.A., Martin, C.H., Smoller, D.A. and Hartl, D.L. (1991) *Proc. Natl Acad. Sci. USA* **88**, 8034.
393. Langerstrom, M., Parik, J., Malmgren, H., Stewart, J., Pettersson, U. and Landegren, U. (1991) *PCR Meth. Appl.* **1**, 111.
394. Rose, E.A. (1991) *FASEB J.* **5**, 46.
406. Straub, R.E., Speer, M.C., Luo, Y., Rojas, K., Overhauser, J., Ott, J. and Gilliam, T.C. (1993) *Genomics* **15**, 48.
407. Porter, J.C., Ram, K.T. and Puck, J.M. (1993) *Genomics* **15**, 57.
408. Kozman, H.M., Phillips, H.A., Callen, D.F., Sutherland, G.R. and Mulley, J.C. (1993) *Cytogenet. Cell Genet.* **62**, 194.
409. Glesne, D., Collart, F., Varkony, T., Drabkin, H. and Huberman, E. (1993) *Genomics* **16**, 274.
410. Chang, E., Welch, S., Luna, J., Giacalone, J. and Franke, U. (1993) *Genomics* **17**, 393.
411. Malo, M.S., Strivastava, K., Andresen, J.M., Chen, X.N., Korenberg, J.R. and Ingram, V.M. (1994) *Proc. Natl Acad. Sci. USA* **91**, 2975.
412. Ha, H., Barnoski, B.L., Sun, L., Emanuel, B.S. and Burrows, P.D. (1994) *J. Immunol.* **152**, 5749.
413. Aslanidis, C. and deJong, P.J. (1991) *Proc. Natl Acad. Sci. USA* **88**, 6765.
414. Guzzetta, V., Montes de Oca Luna, R., Lupski, J.R. and Patel, P.I. (1991) *Genomics* **9**, 31.
415. Bernard, L.E., Brooks-Wilson, A.R. and Wood, S. (1991) *Genomics* **9**, 246.
416. Bicknell, D.C., Markie, D., Spurr, N.K. and Bodmer, W.F. (1991) *Genomics* **10**, 186.
417. Monaco, A.P., Lam, V.M., Zehetner, G., Lennon, G.G.,

395. Rogowsky, P.M., Shepherd, K.W. and Langridge, P. (1992) *Genome* **35**, 621.
396. Telenius, H., Carter, N.P., Bebb, C.E., Nordenskjold, M., Ponder, B.A. and Tunnacliffe, A. (1992) *Genomics* **13**, 718.
397. Butler, R., Ogilvie, D.J., Elvin, P., Riley, J.H., Finniear, R.S., Slynn, G., Morten, J.E., Markham, A.F. and Anand, R. (1992) *Genomics* **12**, 42.
398. Kere, J., Nagaraja, R., Mumm, S., Ciccodicola, A., D'Urso, M. and Schlessinger, D. (1992) *Genomics* **14**, 241.
399. Takahashi, N. and Ko, M.S. (1993) *Genomics* **16**, 161.
400. Silverman, G.A. (1993) *PCR Meth. Appl.* **3**, 141.
401. Petersen, M.B., Weber, J.L., Slaugenhaupt, S.A., Kwitek, A.E., McInnis, M.G., Chakravarti, A. and Antonarakis, S.E. (1991) *Hum. Genet.* **87**, 401.
402. Matsutani, A., Janssen, R., Donis-Keller, H. and Permutt, M.A. (1992) *Genomics* **12**, 319.
403. McNamara, J.O., Eubanks, J.H., McPherson, J.D., Wasmuth, J.J., Evans, G.A. and Heinemann, S.F. (1992) *J. Neurosci.* **12**, 2555.
404. Vandenbergh, D.J., Persico, A.M., Hawkins, A.L., Griffin, C.A., Li, X., Jabs, E.W. and Uhl, G.R. (1992) *Genomics* **14**, 1104.
405. Ahmed, C.M., Ware, D.H., Lee, S.C. *et al.* (1993) *Proc. Natl Acad. Sci. USA* **89**, 8220.

...

Douglas, C., Nizetic, D., Goodfellow, P.N. and Lehrach, H. (1991) *Nucl. Acids Res.* **19**, 3315.
418. Mares, A., Ledbetter, S.A., Ledbetter, D.H., Roberts, R. and Hejmancik, J.F. (1991) *Genomics* **11**, 215.
419. Meese, E.U., Meltzer, P.S., Ferguson, P.W. and Trent, J.M. (1991) *Genomics* **12**, 549.
420. Lengauer, C., Riethman, H.C., Speicher, M.R., Taniwaki, M., Konecki, D., Green, E.D., Becher, R., Olson, M.V. and Cremer, T. (1991) *Cancer Res.* **52**, 2590.
421. Breen, M., Arveiler, B., Murray, I., Gosden, J.R. and Porteous, D.J. (1992) *Genomics* **13**, 726.
422. Lengauer, C., Green, E.D. and Cremer, T. (1992) *Genomics* **13**, 826.
423. Cole, C.G., Dunham, I., Coffey, A.J., Ross, M.T., Meier Ewert, S., Bobrow, M. and Bentley, D.R. (1992) *Genomics* **14**, 256.
424. Dorin, J.R., Emslie, E., Hanratty, D., Farrall, M., Gosden, J. and Porteous, D.J. (1992) *Hum. Mol. Genet.* **1**, 53.
425. Liu, P., Siciliano, J., Seong, D., Craig, J., Zhao, Y., deJong, P.J. and Siciliano, M.J. (1993) *Cancer Genet. Cytogenet.* **65**, 93.
426. Charlieu, J.P., Laurent, A.M., Orti, R., Viegas Pequignot, E., Bellis, M. and Roizes, G. (1993) *Genomics* **15**, 576.
427. Klein, V., Piontek, K., Brass, N., Subke, F., Zang, K.D. and Meese, E. (1993) *Genet. Anal. Tech. Appl.* **10**, 6.

428. Siden, T.S., Kumlien, J., Drumheller, T., Smith, S.E., Rohme, D., Lehrach, H. and Smith, D.I. (1994) *Somat. Cell Mol. Genet.* **20**, 137.
429. Francis, F., Benham, F., See, C.G., Fox, M., Ishikawa Brush, Y., Monaco, A.P., Weiss, B., Rappold, G., Hamvas, R.M. and Lehrach, H. (1994) *Genomics* **20**, 75.
430. Landgraf, A., Reckmann, B. and Pingoud, A. (1991) *Anal. Biochem.* **193**, 231.
431. Lundeberg, J., Wahlberg, J. and Uhlen, M. (1991) *BioTechniques* **10**, 68.
432. Michael, N.L., Vahey, M., Burke, D.S. and Redfield, R.R. (1992) *J. Virol.* **66**, 310.
433. Ferre, F. (1992) *PCR Meth. Appl.* **2**, 1.
434. Porcher, C., Malinge, M.C., Picat, C. and Grandchamp, B. (1992) *BioTechniques* **13**, 106.
435. Clementi, M., Menzo, S., Bagnarelli, P., Manzin, A., Valeza, A. and Veraldo, P.E. (1993) *PCR Meth. Appl.* **2**, 191.
436. Sardelli, A.D. (1991) *Amplifications* **9**, 1.
437. Lee, T., Sunzeri, F.J., Tobler, L.H., Williams, B.G. and Busch, M.P. (1990) *AIDS* **5**, 683.
438. Pang, S., Koyanagi, Y., Miles, S., Wiley, C., Vinters, H.V. and Chen, I.S.Y. (1990) *Nature* **343**, 85.
439. Dickover, R.E., Donovan, R.M., Goldstein, E., Dandekar, S.,

3. Higuchi, R., Krummel, B. and Saiki, R.K. (1988) *Nucl. Acids Res.* **16**, 7351.
4. Jones, P., Qiu, J. and Rickwood, D. (1994) *RNA Isolation and Analysis*. BIOS Scientific Publishers, Oxford.

Chapter 5

1. Saiki, R.K., Gelfand, D.H., Stoffel, S., Scharf, S., Higuchi, R., Horn, G.T., Mullis, K. and Erlich, H. (1988) *Science* **239**, 487.
2. Tindall, K.R. and Kunkel, T.A. (1988) *Biochemistry* **27**, 6008
3. Lawyer, F.C., Stoffel, S., Saiki, R.K., Myambo, K., Drummond, R. and Gelfand, D.H. (1989) *J. Biol. Chem.* **264**, 6427.
4. Lawyer, F.C., Stoffel, S., Saiki, R.K., Chang, S.Y., Landre, P.A., Abramson, R.D. and Gelfand, D.H. (1993) *PCR Meth. Appl.* **2**, 275.
5. Slupphaug, G., Alseth, I., Eftedal, I., Volden, G. and Krokan, H. E. (1993) *Anal. Biochem.* **211**, 164.
6. Myers, T.W. and Gelfand, D.H. (1991) *Biochemistry* **30**, 7661.
7. Katcher, H.L. and Schwartz, I. (1994) *BioTechniques* **16**, 84.
8. de Noronha, C.M.C. and Mullins, J.I. (1992) *PCR Meth. Appl.* **2**, 131.
9. Lundberg, K.S., Schoemaker, D.D., Adams, M.W., Short, J.M., Sorge, J.A. and Mathur, E.J. (1991) *Gene* **108**, 1.

Bush, C.E. and Carlson, J.R. (1990) *J. Clin. Microbiol.* **28**, 2130.
440. Lehtovaara, P., Uusi-Oukari, M., Buchert, P., Laaksonen, M., Bengström, M. and Ranki, M. (1992) *PCR Meth. Appl.* **3**, 169.
441. Noonan, K.E., Beck, C., Holzmayer, J.E. et al. (1990) *Proc. Natl Acad. Sci. USA* **87**, 7160.
442. Uberla, K., Platzer, C., Diamanstein, T. and Blankenstein, T. (1991) *PCR Meth. Appl.* **1**, 136.
443. Hoof, T., Riordan, J.R. and Tummler, B. (1991) *Anal. Biochem.* **196**, 161.
444. Kaneko, S., Murakami, S., Unoura, M. and Kobayashi, K. (1992) *J. Med. Virol.* **37**, 278.
445. Puntschart, A., Jostarndt, K., Hoppeler, H. and Billeter, R. (1994) *PCR Meth. Appl.* **3**, 232.

Chapter 2

1. Kwok, S. and Higuchi, R. (1989) *Nature* **339**, 237.

Chapter 4

1. Chomczynski, P. and Sacchi, N. (1987) *Anal. Biochem.* **162**, 156.
2. Brown, T.A. (ed.) (1991) *Molecular Biology Labfax*. BIOS Scientific Publishers, Oxford.

10. Costa, G.L. and Weiner, M.P. (1994) *Nucl. Acids Res.* **22**, 2423.
11. Knittel, T. and Picard, D. (1993) *PCR Meth. Appl.* **2**, 346.

Chapter 6

1. Nichols, R., Andrews, P.C., Zhang, P. and Bergstrom, D.E. (1994) *Nature* **369**, 492.
2. Liang, P. and Pardee, A.B. (1992) *Science* **257**, 967.
3. Bauer, D., Müller, H., Reich, J., Riedel, H., Ahrenkiel, V., Warthoe, P. and Strauss, M. (1993) *Nucl. Acids Res.* **21**, 4272.
4. Williams, J.G.K., Kubelik, A.R., Livak, K.J., Rafalski, J.A. and Tingey, S.V. (1990) *Nucl. Acids Res.* **18**, 6531.
5. Scharf, S.J., Horn, G.T. and Erlich, H.A. (1986) *Science* **233**, 1076.
6. Sheffield, V.C., Cox, D.R., Lerman, L.S. and Myers, R.M. (1989) *Proc. Natl Acad. Sci. USA* **86**, 232.
7. Stoflet, E.S., Koeberl, D.D., Sarkar, G. and Sommer, S.S. (1988) *Science* **239**, 491.
8. Sarkar, G. and Sommer, S.S. (1989) *Science* **244**, 331.
9. Browning, K.S. (1989) *Amplifications* **3**, 14.
10. Newton, C.R., Graham, A., Heptinstall, L.E., Powell, S.J., Summers, C., Kalsheker, N., Smith, J.C. and Markham, A.F. (1989) *Nucl. Acids Res.* **17**, 2503.

11. Suggs, S.V., Wallace, R.B., Hirose, T., Kawashima, E.H. and Itakura, K. (1981) *Proc. Natl Acad. Sci. USA* **78**, 6613.
12. Sambrook, J., Fritsch, E.F. and Maniatis, T. (eds) (1989) *Molecular Cloning: a Laboratory Manual*, 2nd Edn. Cold Spring Harbor Laboratory Press, Cold Spring Harbor, NY.
13. Wu, D.Y., Ugozzoli, L., Pal, B.K., Qian, J. and Wallace, R.B. (1991) *DNA Cell Biol.* **10**, 233.

Chapter 7

1. Brown, T. and Brown, D.J.S. (1991) in *Oligonucleotides: a Practical Approach* (F. Eckstein, ed.), p. 1. IRL Press, Oxford.
2. Brown, T. and Brown, D.J.S. (1992) in *Meth. Enzymol.* **211**, 20
3. Pon, R. T. (1991) *Tetrahedron Lett.* **32**, 1715.
4. Holtke, H.-J., Seibl, R., Burg, J., Muhlegger, K. and Kessler, C. (1990) *Biol. Chem. Hoppe-Seyler* **371**, 929.
5. Seibl, R., Holtke, H.-J., Ruger, R. *et al.* (1990) *Biol. Chem. Hoppe-Seyler*, **371**, 939.
6. Muhlegger, K., Huber, E., Von Der Eltz, H., Ruger, R. and Kessler, C. (1990) *Biol. Chem. Hoppe-Seyler* **371**, 953.
7. Grzybowski, J., Will, D.W., Randall, R.E., Smith, C.A. and Connolly, B. A. and Rider, P. (1985) *Nucl. Acids Res.* **13**, 4485.
18. Connolly, B. A. and Rider, P. (1985) *Nucl. Acids Res.* **13**, 4485.
19. Sproat, B.S., Beijer, B.S., Rider, P. and Neuner, P. (1987) *Nucl. Acids Res.* **15**, 4837.
20. Zuckerman, R., Corey, D. and Shultz, P. (1987) *Nucl. Acids Res.* **15**, 5305.
21. Dolinnaya. N.G., Blumenfeld, M., Merenkora, I.N., Oretskaya, T.S., Koynetskaya, N.F., Ivanovskaya, M.G., Vasseur, M. and Shabarova, Z.A. (1993) *Nucl. Acids Res.* **21**, 5403.
22. Horn, T. and Urdea, M. (1986) *Tetrahedron Lett.* **27**, 4705.
23. ABI User Bulletin (1989) No. 18. Kelvin Close, Birchwood Science Park, Warrington, UK.

Chapter 8

1. Cobb, B.D. and Clarkson, J.M. (1994) *Nucl. Acids Res.* **22**, 3801.
2. Taguchi, G. (1986) *Introduction to Quality Engineering*. Asian Productivity Organization, UNIPUB, New York.
3. Woodford, K., Wietzmann M.N. and Usdin, K. (1995) *Nucl. Acid Res.* **23**, 539.
4. Wittwer, C.T. and Garling, D.J. (1991) *BioTechniques* **10**, 76.
5. Rychlik, W., Spencer, W.J. and Rhoads, R.E. (1990) *Nucl. Acids Res.* **18**, 6409.

References

7. Brown, T. (1993) *Nucl. Acids Res.* **21**, 1705.
8. Urdea, M.S., Warner, B.D., Running, J.A., Stempien, M., Clyne, J. and Horn, T. (1988) *Nucl. Acids Res.* **16**, 4937.
9. *ABI User Bulletin* (1989) No. 11. Kelvin Close, Birchwood Science Park, Warrington, UK.
10. Schubert, F., Ahlert, K., Cech, D. and Rosenthal, A. (1990) *Nucl. Acids Res.* **18**, 3427.
11. Smith, L.M., Sanders, J.Z., Kaiser, R.J., Hughes, P., Dod, C., Connell, C.R., Heiner, C., Kent, S.B.H. and Hood, L.E. (1986) *Nature* **321**, 674.
12. Argawal, S. and Zamecnik, P.C. (1990) *Nucl. Acids Res.* **18**, 5419.
13. Nichols, R., Andrews, P.C., Zhang, P. and Bergstrom, D.E. (1994) *Nature* **369**, 492.
14. Newton, C. R., Holland, D., Heptinstall, L. E., Hodgson, I., Edge, M.D., Markham, A.F. and McLean, M.J. (1993) *Nucl. Acids Res.* **21**, 1155.
15. Gade, R., Kaplan, B.E., Swiderski, P.M. and Wallace, R.B. (1993) *Genet. Anal. Techn. Appl.* **10**, 61.
16. Alves, A.M., Edge, M.D. and Holland, D. (1989) *Tetrahedron Lett.* **30**, 3089.
17. Misiura, K., Durrant, I., Evans, M.R. and Gait, M.J. (1990) *Nucl. Acids Res.* **18**, 4345.
6. Cheng, S., Fockler, C., Barnes, W.M. and Higuchi, R. (1994) *Proc. Natl. Acad. Sci. USA* **91**, 5695.
7. Lindahl, T. and Nyberg, B. (1972) *Biochem.* **11**, 3610.
8. Barnes, W. (1994) *Proc Natl Acad. Sci. USA* **91**, 2216.
9. Ponce, M.R. and Micol, J.L. (1992) *Nucl. Acids Res.* **20**, 623.
10. Winship, P.R. (1989) *Nucl. Acids Res.* **17**, 1266.
11. Pomp, D. and Medrano, J.F. (1991) *BioTechniques* **10**, 58.
12. Sun, Y., Hegamyer, G. and Colburn, N.H. (1993) *BioTechniques* **15**, 372.
13. Filichkin, S.A. and Gevin, S.B. (1992) *BioTechniques* **12**, 828.
14. Varadaraj, K. and Skinner, D.M. (1994) *Gene* **140**, 1.
15. Berger, S.L. (1994) *Anal. Biochem.* **222**, 290.
16. Shen, W.-H. and Hohn, B. (1992) *Trends Genet.* **8**, 227.
17. Lu, Y.-H. and Nègre, S. (1993) *Trends Genet.* **9**, 297.
18. Hung, T., Mak, K. and Fong, K. (1990) *Nucl. Acids Res.* **18**, 4953.
19. Sarkar, G., Kapelner, S. and Sommer, S.S. (1990) *Nucl. Acids Res.* **18**, 7465.
20. Gelfand, D.H. (1989) in *PCR Technology: Principles and Applications for DNA Amplification* (H.A. Erlich, ed.), p. 17. Stockton Press, New York.
21. Comey, T.C., Jung, J.M. and Budowle, B. (1991) *BioTechniques* **10**, 60.

22. Newton, C.R. and Graham, A. (1994) *PCR*. BIOS Scientific Publishers, Oxford.
23. Neilson, K. and Mathur, E.J. (1990) *Stratagies* **3**, 17.
24. Schwarz, K., Halsen-Hagge, T. and Bartram, B. (1990) *Nucl. Acids Res.* **18**, 1079.

Chapter 9

1. Victor, T., Jordaan, A., du Toit, R. and Van Helden, P.D. (1993) *Eur. J. Clin. Chem. Clin. Biochem.* **31**, 531.
2. Uo, C.-Y., Moore, J.L. and Schochetman (1991) *BioTechniques* **10**, 442.
3. Sarkar, G. and Sommer, S.S. (1991) *BioTechniques* **10**, 591.
4. Fox, J.C., Ait-Khaled, M., Webster, A. and Emery, V.C. (1991) *J. Virol. Meth.* **33**, 375.
5. Deragon, J.M., Sinnett, D., Mitchell, G., Potier, M. and Labuda, D. (1990) *Nucl. Acids Res.* **18**, 6149.
6. Furrer, B., Candrian, U., Wieland, P. and Lüthy, J. (1990) *Nature* **346**, 324.
7. DeFilippes, M.M. (1991) *BioTechniques* **10**, 26.
8. Strom, C.M., Rechitsky, S. and Verlinsky, Y. (1991) *J. In Vitro. Fert. Embryo Transfer* **8**, 225.
22. Longo, M.C., Berninger, M.S. and Hartley, J.L. (1990) *Gene* **93**, 125.
23. Walder, R.Y., Hayes, J.R. and Walder, J.A. (1993) *Nucl. Acids Res.* **21**, 4339.
24. Mivechi, N.F. and Rossi, J.J. (1990) *Cancer Res.* **50**, 2877.
25. Shuldiner, A.R., Tanner, A.R., Moore, C.A. and Roth, J. (1991) *BioTechniques* **11**, 760.
26. Grillo, M. and Margolis, F.L. (1990) *BioTechniques* **9**, 262.
27. Dilworth, D.D. and McCarrey, J. (1992) *PCR Meth. Appl.* **1**, 279.
28. Dougherty, R.M., Phillips, P.E., Gibson, S. and Young, L. (1993) *J. Virol. Meth.* **41**, 235.
29. Pang, J., Modlin, J. and Yolken, R. (1992) *Mol. Cell. Probes* **6**, 251.
30. Udaykumar, Epstein, J.S. and Hewlett, I.K. (1993) *Nucl. Acids Res.* **21**, 3917.
31. Cone, R.W. and Fairfax, M.R. (1993) *PCR Meth. Appl.* **3**, S15.
32. Prince, A.M. and Andrus, L. (1992) *BioTechniques* **12**, 358.

Chapter 10

1. Sanderson, J.D., Moss, M. T., Tizard, M. L. V. and Hermon-

9. Zhu, Y.S., Isaacs, S.T., Cimino, G.D. and Hearst, J.E. (1991) *Nucl. Acids Res.* **19**, 2511.
10. Muralidhar, B. and Steinman, C.R. (1992) *Gene* **117**, 107.
11. Cimino, G.D., Metchette, K.C., Tessman, J.W., Hearst, J.E. and Isaacs, S.T. (1991) *Nucl. Acids Res.* **19**, 99.
12. Isaacs, S.T., Tessman, J.W., Metchette, K.C., Hearst, J.E. and Cimino, G.D. (1991) *Nucl. Acids Res.* **19**, 109.
13. Espy, M.J., Smith, T.F. and Persing, D.H. (1993) *J. Clin. Microbiol.* **31**, 2361.
14. Rys, P.N. and Persing, D.H. (1993) *J. Clin. Microbiol.* **31**, 2356.
15. Jinno, Y., Yoshiura, K. and Niikawa, N. (1990) *Nucl. Acids Res.* **18**, 6739.
16. Meier, A., Persing, D.H., Finken, M. and Böttger, E.C. (1993) *J. Clin. Microbiol.* **31**, 646.
17. Aslanzadeh, J. (1993) *Mol. Cell. Probes* **7**, 145.
18. Wang, X., Chen, T., Kim, D. and Piomelli, S. (1992) *Am. J. Hematol.* **40**, 146.
19. Glenn, T.C., Waller, D.R. and Braun, M.J. (1994) *BioTechniques* **17**, 1086.
20. Slupphaug, G., Alseth, I., Eftedal, I., Volden, G. and Krokan, H.E. (1993) *Anal. Biochem.* **211**, 164.
21. Thornton, C.G., Hartley, J.L. and Rashtchian, A. (1992) *BioTechniques* **13**, 180.
22. Taylor, J. (1992) *Gut* **33**, 890.
23. Challans, J., Stevenson, K., Reid, H.W. and Sharp, J.M. (1994) *Vet. Rec.* **134**, 95.
24. Grzybowski, J., Will, D.W., Randall, R.E., Smith, C.A. and Brown, T. (1993) *Nucl. Acids Res.* **21**, 1705.
25. MacKellar, C., Graham, D., Will, D.W., Burgess, S. and Brown, T. (1992) *Nucl. Acids Res.*, **20**, 3411.

Chapter 11

1. Higuchi, R., Dollinger, G., Walsh, P.S. and Griffith, R. (1992) *BioTechnology* **10**, 413.
2. Higuchi, R., Fockler, C., Dollinger, G. and Watson, R. (1993) *BioTechnology* **11**, 1026.
3. Sharp, P.A., Sugden, B. and Sambrook, J. (1973) *Biochemistry* **12**, 3055.
4. Merril, C.R. (1990) *Meth. Enzymol.* **182**, 477.
5. Bassam, B.J., Caetano-Anollés, G. and Gresshoff, P.M. (1991) *Anal. Biochem.* **196**, 81.
6. Bassam, B.J. and Caetano-Anollés, G. (1993) *Appl. Biochem. Biotechnol.* **42**, 181.
7. Holland, P.M., Abramson, R.D., Watson, R. and Gelfand, D.H. (1991) *Proc. Natl Acad. Sci. USA* **88**, 7276.

8. Holland, P.M., Abramson, R.D., Watson, R., Will, S., Saiki, R.K. and Gelfand, D.H. (1992) *Clin. Chem.* **38**, 462.
9. Lee, L.G., Connell, C.R. and Bloch, W. (1993) *Nucl. Acids Res.* **21**, 3761.
10. Kaluz, S. and Reid, K.B.M. (1991) *Nucl. Acids Res.* **19**, 4012.
11. Patel, D. (1994) *Gel Electrophoresis: Essential Data.* John Wiley & Sons, Chichester.
12. Sambrook, J., Fritsch, E.F. and Maniatis, T. (eds) (1989) *Molecular Cloning: a Laboratory Manual,* 2nd Edn. Cold Spring Harbor Laboratory Press, Cold Spring Harbor, NY.
13. Fodde, R. and Losekoot, M. (1994) *Hum. Mutat.* **3**, 83.
14. Fischer, S.G. and Lerman, L.S. (1980) *Proc. Natl Acad. Sci. USA* **77**, 4420.
15. Fischer, S.G. and Lerman, L.S. (1983) *Proc. Natl Acad. Sci. USA* **80**, 1579.
16. Myers, R.M., Maniatis, T. and Lerman, L.S. (1987) *Meth. Enzymol.* **155**, 501.
17. Myers, R.M., Sheffield, V.C. and Cox, D.R. (1988) in *Genome Analysis: a Practical Approach* (K. Davies, ed.). Oxford University Press, Oxford.
18. Feinberg, A.P. and Vogelstein, B. (1983) *Anal. Biochem.* **132**, 6.
19. Feinberg, A.P. and Vogelstein, B. (1984) *Anal. Biochem.* **137**, 266.
32. Cuppens, H., Buyse, I., Baens, M., Marynen, P. and Cassiman, J. (1992) *Mol. Cell. Probes* **6**, 33.
33. Murray, V. (1989) *Nucl. Acids Res.* **17**, 8889.
34. Galas, D.J. and Schmitz, A. (1978) *Nucl. Acids Res.* **5**, 3157.
35. Zicetkiewicz, E., Rafalski, A. and Labuda, D. (1994) *Genomics* **20**, 176.
36. Liang, P. and Pardee, A.B. (1992) *Science* **257**, 967.
37. Bauer, D., Müller, H., Reich, J., Riedel, H., Ahrenkiel, V., Warthoe, P. and Strauss, M. (1993) *Nucl. Acids Res.* **21**, 4272.
38. Spann, W., Pachmann, K., Zabnienska, H, Pielmeier, A. and Emmerich, B. (1991) *Infection* **19**, 242.
39. Bagasra, O., Seshamma, T. and Pomerantz, R.J. (1993) *J. Immunol. Meth.* **158**, 131.
40. Cotton, R.G.H., Rodrigues, N.R. and Campbell, R.D. (1988) *Proc. Natl Acad. Sci. USA* **85**, 4397.
41. Vary, C.P.H. (1992) *Clin. Chem.* **38**, 687.
42. Kemp, D.J., Smith, D.B., Foote, S.J., Samaras, N. and Peterson, M.G. (1989) *Proc. Natl Acad. Sci. USA* **86**, 2423.
43. Mayrand, P.E., Corcoran, K.P., Ziegle, J., Robertson, J., Hoff, L.B. and Kronick, M. (1992) *Appl. Theor. Electrophor.* **3**, 1.

20. Kricka, L.J. (1992) *Nonisotopic DNA Probe Techniques*. Academic Press, New York.
21. Southern, E.M. (1975) *J. Mol. Biol.* **98**, 503.
22. Saiki, R.K., Bugawan, T.L., Horn, G.T., Mullis, K.B. and Erlich, H.A. (1986) *Nature* **324**, 163.
23. Bugawan, T.L., Saiki, R.K., Levenson, C.H., Watson, R.M. and Erlich, H.A. (1988) *BioTechnology* **6**, 943.
24. Saiki, R.K., Chang, C.-A., Levenson, C.H., Warren, T.C., Boehm, C.D., Kazazian, H.H. and Erlich, H.A. (1988) *N. Engl. J. Med.* **319**, 537.
25. Houlston, R.S., Snowden, C., Green, F., Alberti, K.G.M.M. and Humphries, S.E. (1989) *Hum. Genet.* **83**, 364.
26. Bugawan, T.L., Begovich, A.B. and Erlich, H.A. (1990) *Immunogenet.* **32**, 231.
27. Bugawan, T.L., Begovich, A.B. and Erlich, H.A. (1991) *Immunogenet.* **34**, 413.
28. Saiki, R.K., Walsh, P.S., Levenson, C.H. and Erlich, H.A. (1989) *Proc. Natl Acad. Sci. USA* **86**, 6230.
29. Zhang, Y., Coyne, M.Y., Will, S.G., Levenson, C.H. and Kawasaki, E.S. (1991) *Nucl. Acids Res.* **19**, 3929.
30. Serre, J.L., Taillandier, A., Mornet, E., Simon-Bouy, B., Boué, J. and Boué, A. (1991) *Genomics* **11**, 1149.
31. Chehab, F.F. and Wall, J. (1992) *Hum. Genet.* **89**, 163.

Chapter 12

1. Rickwood, D. and Hames, B.D. (eds) (1990) *Gel Electrophoresis of Nucleic Acids: a Practical Approach*, 2nd Edn. Oxford University Press, Oxford.
2. Sambrook, J., Fritsch, E.F. and Maniatis, T. (eds) (1989) *Molecular Cloning: a Laboratory Manual*, 2nd Edn. Cold Spring Harbor Laboratory Press, Cold Spring Harbor, NY.
3. Krowczynska, A.M. and Henderson, M.B. (1992) *BioTechniques* **13**, 286.

Chapter 13

1. Marchuk, D., Drumm, M., Saulino, A. and Collins, F.S. (1991) *Nucl. Acids Res.* **19**, 1154
2. Clarke, J.M. (1988) *Nucl. Acids Res.* **16**, 9677.
3. Liu, Z. and Schwartz, L.M. (1992) *BioTechniques* **12**, 28.
4. Aslanidis, C. and deJong, P.J. (1990) *Nucl. Acids Res.* **18**, 6069.
5. Haun, R.S., Serventi, I.M. and Moss, J. (1992) *BioTechniques* **13**, 515.
6. Haun. R.S. and Moss, J. (1992) *Gene* **112**, 37.
7. Nisson P.E., Rashtchian, A. and Watkins, P.C. (1991) *PCR Meth. Appl.* **1**, 120.

8. Rashtchian, A., Buchman, G. W., Schuster, D. M. and Berninger, M. (1991) *Anal. Biochem.* **206**, 91.
9. Scharf, S.J., Horn, G.T. and Erlich, H.A. (1986) *Science* **233**, 1076.
10. Clackson, T.P. and Winter, G. (1989) *Nucl. Acids Res.* **17**, 10163.
11. Kaufman, D.L. and Evans, G.A. (1990) *BioTechniques* **9**, 304.

Chapter 14

1. Bennet, B. and Molenaar, A. (1994) *BioTechniques* **16**, 32.
2. Ho, S.F., Pullen, J.K., Horton, R.M., Hunt, H.D. and Pease, L.R. (1989) *DNA Prot. Eng. Tech.* **2**, 50.

Chapter 15

1. Barnes, W.M. (1994) *Proc. Natl Acad. Sci. USA* **91**, 2216.
2. Costa, G.L. and Weiner, M.P. (1994) *Nucl. Acids Res.* **22**, 2423.
3. Jung, V., Pestka, S.B. and Pestka, S. (1990) *Nucl. Acids Res.* **18**, 6156.

FURTHER READING

1 Companion volume to present book

Newton, C.R. and Graham, A. (1994) *PCR*. BIOS Scientific Publishers, Oxford.

2 Other books

Davies, K. (ed.) (1988) *Genome Analysis: a Practical Approach*. Oxford University Press, Oxford.
Ellingboe, J. and Gyllensten, U.B. (eds) (1992) *The PCR Technique: DNA Sequencing*. Eaton Publishing, Natick, MA.
Erlich, H.A. (ed.) (1989) *PCR Technology*. Stockton Press, New York.
Griffin, H.G. and Griffin, A.M. (eds) (1994) *PCR Technology: Current Innovations*. CRC Press, Boca Raton, FL.
Innis, M.A., Gelfand, D.H., Sninsky, J.J. and White, T.J. (eds) (1990) *PCR Protocols: a Guide to Methods and Applications*. Academic Press, San Diego.
Jones, P., Qui, J. and Rickwood, D. (1994) *RNA Isolation and Analysis*. BIOS Scientific Publishers, Oxford.
McPherson, M.J., Hames, B.D. and Taylor, G.R. (eds) (1995) *PCR-II: a Practical Approach*. Oxford University Press, Oxford.
McPherson, M.J., Quirke, P. and Taylor, G.R. (eds) (1991) *PCR: a Practical Approach*. Oxford University Press, Oxford.
Mullis, K.B., Ferré, F. and Gibbs, R.A. (eds) (1994) *The Polymerase Chain Reaction*. Birkhäuser, Boston.
Rolfs, A., Schuller, I., Finckh, U. and Weber-Rolfs, I. (1992) *PCR: Clinical Diagnostics and Research*. Springer-Verlag, Berlin.
White, B.A. (ed.) (1993) *Methods in Molecular Biology 15*.

PCR Protocols; Current Methods and Applications. Humana Press, Totowa, NJ.

3 Journals and newsletters

Amplifications. A forum for PCR users. A news-sheet produced by Perkin-Elmer; contact Applied Biosystems.

BioTechniques. Eaton Publishing, Natick, MA.
Nucleic Acids Research. Oxford University Press, Oxford.
PCR Methods and Applications. Cold Spring Harbor Laboratory Press, Cold Spring Harbor, NY.
Stratagies. A quarterly newsletter produced by Stratagene.

APPENDIX

Trade names and registered trademarks

Trade names and registered trademarks are the properties of the companies shown in *Table 1*.

Table 1. Trade names and registered trademarks

Product	Company	Product	Company
AMPLICOR™	Roche Diagnostic Systems	CloneAmp®	Life Technologies
Amplifications®	Perkin-Elmer	Crocodile™	Appligene
AmpliTaq®	Hoffmann-La Roche	CsTFA™	Pharmacia
AmpliType®	Hoffmann-La Roche	Cyclist™	Stratagene
AmpliWax™	Hoffmann-La Roche	Deep Vent®	New England Biolabs
ARMS™	Zeneca	DEVIATS™	Zeneca
Bio-Dot®	Bio Rad	DNA DipStick™	Invitrogen
Biodyne®	Life Technologies	Duralon-UV™	Stratagene
Captagene™	Pharmacia	Duralose-UV™	Stratagene
Centricon®	Amicon	Dynabeads®	Dynal
CHROMA SPIN™	Clontech	EASIgel™	Scotlab
CircumVent™	Life Technologies	*ECL*™	Amersham

Continued

Table 1. Trade names and registered trademarks, *continued*

Product	Company	Product	Company
EnviroAmp™	Hoffmann–La Roche	PhastSystem™	Pharmacia
FastCheck™	Life Technologies	PolyATact®	Promega
Ficoll®	Pharmacia	Positive™	Appligene
fmol®	Promega	PowerBlock™	Ericomp
Prep-A-Gene®	Bio Rad	PrimeErase® Quik™	Stratgene
FPLC®	Pharmacia	PTC-100™	M J Research
GeneAmp®	Hoffmann–La Roche	PTC-150™	M J Research
GeneAmp™	Hoffmann–La Roche	pTOPE™	Novagen
GeneAmplimer®	Hoffmann–La Roche	QIAGEN®	Qiagen
Geneclean®	Bio 101	QIAquick™	Qiagen
GENE-TECH®	Stuart Scientific	QPCR™	Perkin-Elmer
GEN-ETI-K®	Sorin Biomedica	*Quant*AMP™	Amersham
GlassMAX™	Life Technologies	RNAimage™	GenHunter
GTG®	FMC Corporation	RNAmap™	GenHunter
Horizon®	Life Technologies	RoboCycler®	Stratagene
Hot Tub™	Amersham	Sephadex®	Pharmacia
Hybond™	Amersham	SHARP Signal™	Digene Diagnostics
IsoGene™	Perkin-Elmer	Sigmacote®	Sigma
MagneSphere®	Promega	SILVER SEQUENCE™	Promega
Magpie™	NBL Gene Sciences	SingleBlock™	Ericomp
Marathon™	Clontech	Single dA™	Novagen
Mastercycler®	Eppendorf	SingleStar™	Ericomp
MicroAmp™	Hoffmann–La Roche	Spinbind™	FMC BioProducts

Microcon™	Amicon	SSAM™	Clontech
MicroSpin™	Pharmacia	T7.Tag™	Novagen
MINI-GENE®	Stuart Scientific	TA Cloning®	Invitrogen
NENSORB®	DuPont NEN	TaqStart®	Clontech
Neutral™	Appligene	The Imager™	Appligene
NucleiClean™	Sigma	Thermalase Tbr™	Amresco
OIL AWAY™	Bio 101	Trans-Blot®	BioRad
Opti-Prime™	Stratagene	Triton™	Rohm and Haas
Parafilm®	Lindsay & Williams Industries	TwinBlock™	Ericomp
pCITE™	Novagen	Tween™	Rohm and Haas
pCR™	Invitrogen	TwinStar™	Ericomp
PCR-Direct™	Clontech	UITma™	Hoffmann-La Roche
PCR MIMICs™	Clontech	UltraPure™	Life Technologies
PCR Optimizer™	Invitrogen	VacuGene™	Pharmacia
PCR SELECT®	5'→3'	Vectorette™	Zeneca
pCR-Script™	Stratagene	Vent®	New England Biolabs
Perfect Match®	Stratagene	VisiGel™	Stratagene
Personal Cycler™	Biometra	Wizard™	Promega
pGEM®	Promega	Zeta-Probe®	Bio Rad

Appendix

INDEX

Agarose gel, 100, 102–104, 110, 115, 118–120, 128, 130, 131
Alkali treatment, 87
Amplicon, 7, 39, 49, 73, 77, 87–92, 103, 132, 167
   capture, 69, 94
   detection, 68–70, 93, 94, 99–125
   labeling, 102, 109
   purification, 128–131
   see also PCR product
Amplimer, 1, 8, 9, 53, 54, 77, 88, 128, 129, 143, 144, 164, 165, 167
   see also Primer
Annealing, 1, 12, 76
   temperature, 50, 51, 85, 91, 164, 166
ARMS, 40, 50, 55, 80
   multiplex, 55

directional, 71, 132, 142, 144
expression, 4
ligation-independent, 139, 142
TA, 4, 39, 40, 141, 166
UDG, 139
Consumables, 10
Contamination
   avoidance, 4, 7, 55, 87–92
   carry-over, 87, 90, 100, 143, 165
   cross, 8, 24
   see also PCR product carry-over
Controls, 9, 11
COP, 55

dATP, 37, 155
dCTP, 37, 155
7 Deaza dGTP, 82, 163

fingerprinting, 6, 80
genomic 27, 75, 77, 87, 90, 91, 105
immobilization, 105
oxidation, 92
polymerase, 9, 37–48, 58, 74, 75, 77, 78, 81, 82, 128, 129, 166, 167
   Deep Vent, 38, 39, 45, 47, 48
   Pfu, 38, 44, 45, 47, 48, 83, 86, 165, 167
   Pfu (exo⁻), 38, 39, 47, 48, 83, 86
   Psp, 44, 45, 47, 48, 83, 86
   Psp (exo⁻), 38, 39, 47, 48, 83, 86
   Pwo, 44, 45, 47, 48
   Stoffel fragment, 40, 41, 47, 48, 83
   Taq, 3, 39, 40, 47, 48, 68, 69, 78, 81, 83, 86, 140, 143, 144, 161–163

Index

ATP, 54, 155
Avidin, 59, 93

Biotin, 58, 59, 68, 70, 93, 95–97, 106, 112, 113, 128
   long chain, 95–97
Blot, 104, 123–126
   dot, 104, 107, 111, 123, 125, 126
   reverse dot, 104, 106, 107, 111, 123, 125, 126
   Southern, 106, 110, 111, 123, 125–127
5-Bromodeoxyuridine, 66, 71

cDNA synthesis, 26, 32, 42, 80
Cholesterol, 97
Chromosome localization, 27
Cloning, 4, 128, 132–142
   blunt-end ligation, 132

Denaturation, 12, 75, 81, 105
   temperature, 85
Deoxyinosine, 45, 49, 66–68, 165
Detection
   chemiluminescent, 93, 106
   colorimetric, 93
   sensitivity, 94, 95
DGGE, 6, 50, 55, 105, 110, 122
dGTP, 37, 155, 163
Diagnosis
   genetic, 5
   preimplantation, 6, 55
Digoxigenin, 57, 59–61, 65, 70, 93, 94, 97, 106, 112
Dinitrophenyl, 61, 62, 70, 93–95, 97
DMSO, 78, 82, 163
DNA
   damage, 78
   extraction, 24, 25, 31

Tbr, 40, 41, 47, 48, 78, 83
Tfl, 41, 47, 48, 84
Tli, 45, 47, 48, 84, 86
Tli (exo⁻), 41, 42, 47, 48, 78, 84
Tma, 46–48, 84, 86
Tth, 42, 43, 45, 47, 48, 84, 86, 162
Vent, 41, 42, 47, 48
sequencer, 93, 108, 111
sequencing, 4, 5, 101, 106, 111, 114, 128
substrate, 1
synthesis, 37
template, 9, 75
DNase, 87
DNP – *see* Dinitrophenyl
dNTPs, 1, 3, 37, 58, 73, 81, 106, 108, 128, 129, 143, 144, 165, 167
dTTP, 37, 42, 44–46, 87, 89, 91, 155
dUMP, 40, 42, 49

213

dUTP, 38, 42, 44–46, 87, 89, 91, 165

Electrophoresis buffers, 116, 129
EtBr, 89, 100, 102–105, 109, 110, 112, 115, 117, 123, 129, 165
Ethidium bromide – *see* EtBr
Exonuclease activity, 37–46, 48, 55, 78, 83, 84, 87, 167

Fluorescein, 64, 65, 68–70, 93, 94, 96, 98, 106
Fluorescence, 100, 115
 laser-induced, 93, 109, 111
Fluorescent dyes, 57, 61–65, 69, 70, 93
Formamide, 53, 82, 163

Gamma irradiation, 87, 88
GAWTS, 50, 55
GC
 clamp, 50

Nitrocellulose, 104, 105, 124
Nonidet, 82
Nucleotide analogs, 49
Nylon, 104, 105, 124

PCR
 alu-, 6, 80
 arbitrarily primed, 6
 asymmetric, 4
 buffers, 73, 74, 81, 143, 147
 DDRT-, 4, 50, 55, 106
 decontamination, 88
 efficiency, 27, 74, 94, 95, 98
 enhancers, 74, 82
 expression, 4, 50
 fidelity, 79, 83–85
 inhibitors, 25
 in situ, 5, 12, 106
 in situ RT-, 4

 quantitative, 6, 26
 specificity, 49, 51, 75–79, 83–85, 100, 111
 stopper, 67–69, 71
 vectorette, 5
 yield, 72, 79, 81, 83–85, 99, 101
PEG, 82
Plasmid, 75
Polyacrylamide gel, 100, 103, 104, 110–112, 115, 118–121
Primer, 1, 3, 37, 49–56, 75, 93, 94, 133, 134, 166
 amino modification, 59, 61, 63, 70, 94
 binding, 85
 chemically labeled, 93–98
 concentration, 3, 52, 81
 degeneracy, 49, 50
 degradation, 37, 43, 44

Index

content, 51
Glycerol, 78, 82, 165

Histological specimens, 5
Hot-start, 4, 44, 75, 76, 164, 166
Hybridization, 101, 105–107, 111, 123–125
Hydroxylamine hydrochloride, 87, 89

5-Iododeoxyuridine, 66
Isopsoralen, 87, 89

Melting temperature, 6, 51–53, 76, 106, 107
Microsatellite, 5, 106
mRNA extraction, 28, 29
Multiple cloning sites, 132–138
Mutagenesis, 4, 51, 55, 80
   insertion, 132

inverse, 5
ligation-anchored, 4
long, 4, 46, 77, 78, 85
machine – *see* Thermal cycler
misincorporation, 39
multiplex, 4, 41, 55, 107, 108
nested, 4, 76, 77
optimization, 73–86
preparative, 72
product
   carry-over, 7, 38, 40, 45, 46, 87, 143, 167
   detection, 57, 93, 99–126
   polishing, 44, 166, 167
   see also Amplicon
recombinant, 4, 55
RNA – *see* PCR, RT-
RT-, 4, 26, 42, 43, 87, 88, 90, 162
RT *in situ*, 5

design, 49–51, 55, 75
dimer, 51, 75, 100, 128
extinction coefficients, 50
labeling, 53–54, 106
synthesis, 44, 54, 57–71
see also Amplimer
Probe, 105–107, 111, 123, 128
Promoter, 132–138
Proofreading, 37
Psoralen, 87, 89

RACE, 4, 27
Radioisotopes, 145, 155–157
Radiolabel, 93
Ramp time, 12–15
RAPD, 6, 41, 50, 55, 80
RAWTS, 50, 55
Reporter, 93, 94

Restriction
enzyme, 87, 88, 91, 101, 110, 122, 128, 132, 133, 141–143, 145, 147, 154, 167
site, 6, 55
Reverse transcriptase, 43, 48
Rhodamine, 65, 70, 93, 106
RNA, 88
cellular, 42
extraction, 24, 26, 28–30
messenger – *see* mRNA
viral, 42
RPFL, 6, 110, 111

Sequencing, 4, 80

Silver staining, 103–105, 110, 111
Sodium hypochlorite, 87, 92
SSCP, 6
Streptavidin, 59, 93, 113
Substrates, 25–36

Target, 1–3, 25, 27, 49, 88, 89, 99, 163
Template, 1, 3, 24, 27, 37, 72, 73, 76, 77, 81, 89, 163, 166
Terminal transferase, 58
Tetraethylene glycol, 64, 65, 71, 94, 95
Texas red, 93
Thermal cyclers, 12–23, 77, 85, 100, 165

Thermal cycling, 1, 12, 74–77, 85, 100, 108
TMAC, 82
TTP – see dTTP
Tween, 82

Uracil DNA glycosylase, 38, 42, 45, 46, 66, 87–91, 139, 140, 165
UV irradiation, 87–89, 92, 100, 115, 129

Vector, 132–141, 167
TA, 4, 43, 132–136

Wax beads, 10, 76

At the forefront of the field...

CURRENT PROTOCOLS IN MOLECULAR BIOLOGY

Edited by F.M. Ausubel, R. Brent, R.E. Kingston, D.D. Moore, Massachusetts General Hospital, Boston, USA, J.G. Seidman, K. Struhl, Harvard Medical School, Boston, USA, and J.A. Smith, University of Alabama, USA

Endorsement: *'This is a thoroughly satisfying compilation, very user friendly and well laid out... it has been well edited and great care has been given to presentation.'*
Nature

Carefully reviewed protocols from over 125 contributors are evaluated by the seven-member editorial board providing the most current information in molecular biology in looseleaf format and on CD-ROM. Quarterly updates contain current advances and new techniques that have been tried and tested in leading labs. The CPMB looseleaf service features over 2,200 pages with hundreds of 'lab-tested' basic, alternate and support protocols with quarterly updates. Subscribers to CPMB on CD-ROM receive the same information plus the added benefits of computer-assisted research capabilities.

For further information please contact:

John Wiley & Sons Inc., 605 Third Avenue, New York, NY 10158-0012, USA

Books from Wiley...

PCR: Essential Techniques

by Julian Burke, University of Sussex, UK

One of the first books in the *Essential Techniques Series*, which contains the most commonly applied techniques in a clear and concise manner allowing quick and easy access to the key protocols used by the life scientist in the lab. These handy pocket-sized manuals are readily transportable, conveniently spiral-bound to lie flat on the lab bench and include useful practical tips. Consisting of a similar number of protocols to larger books, *PCR: Essential Techniques* provides value for money by giving accurate up-to-date quality information in a single source.

0471 95697 X   160pp   Nov 1995   (pr)   £14.95

Short Protocols In Molecular Biology 3rd Edn

Edited by F.M. Ausubel, R. Brent, R.E. Kingston, D.D. Moore, J.G. Seidman, K. Struhl, and J.A. Smith

The third edition of this laboratory manual guides researchers through experimental procedures and methods in molecular biology and provides a companion to the 3-volume *Current Protocols in Molecular Biology*. Methods are easy to follow in a step-by-step format and are all 'lab-tested', contributed and carefully edited by leading international laboratories.

0471 13781 2   approx 850pp   Nov 1995   (pr) $74.95

Please contact Wiley for further information.

ESSENTIAL DATA SERIES

All researchers need rapid access to data on a daily basis. The *Essential Data* series provides this core information in convenient pocket-sized books. For each title, the data have been carefully chosen, checked and organized by an expert in the subject area. *Essential Data* books therefore provide the information that researchers need in the form in which they need it.

Centrifugation
D. Rickwood, T.C. Ford & J. Steensgaard
0 471 94271 5, March 1994, £12.95/$19.95

Gel Electrophoresis
D. Patel
0 471 94306 1, March 1994, £12.95/$19.95

Light Microscopy
C. Rubbi
0 471 94270 7, April 1994, £12.95/$19.95

Vectors
P. Gacesa & D. Ramji
0 471 94841 1, September 1994, £12.95/$19.95

Human Cytogenetics
D. Rooney & B. Czepulkowski (Eds)
0 471 95076 9, October 1994, £12.95/$19.95

Animal Cells: culture and media
D.C. Darling & S.J. Morgan
0 471 94300 2, November 1994, £12.95/$19.95

Cell and Molecular Biology
D. Rickwood & D. Patel
0 471 95568 X, May 1995, £14.95/$23.95

PCR
C.R. Newton (Ed.)
0 471 95222 2, June 1995, £14.95/$23.95

ORDER FORM

Please send me:

Qty	Title	Price/copy	Total
.....	....	....	....
.....	....	....	....
.....	....	....	....

All prices correct at time of going to press but subject to change. Your order will be processed without delay, please allow 21 days for delivery. We will refund your payment without question if you return any unwanted book to us in resaleable condition within 30 days. All books are available from your bookseller.

Method of payment

☐ Payment £/$ _____ enclosed (payable to John Wiley & Sons Ltd). Orders for one book only – please add £3.00/$4.95 to cover postage and handling. Two or more books postage FREE.

☐ Purchase order enclosed
☐ Please send me an invoice (£3.00/$4.95 will be added to cover postage and handling)
☐ Please charge my credit card account
  ☐ American Express ☐ Diners Club
  ☐ Visa ☐ Mastercard

Card no. _____ Expiry: _____

Signature:

Telephone our Customer Services Dept with your cash or credit card order on 01243 829121 or dial FREE on 0800 243407 (UK only)

Send my order to:

Name (PLEASE PRINT) _____

Position: _____

Address: _____

Telephone: _____

Signature: _____ Date: _____

Return to: Rebecca Harfield, John Wiley & Sons Ltd, Baffins Lane, Chichester, West Sussex PO19 1UD, UK. Telefax: (01243) 770225

or: Wiley-Liss, 605 Third Avenue, New York, NY 10158-0012, USA. Telefax: (212) 850-8888

☐ If you do not wish to receive mailings from other companies please tick this box or notify the Marketing Services Department at John Wiley & Sons Ltd.

WILEY